Units and conversion charts

a handbook
for engineers and scientists

Units and conversion charts

a handbook
for engineers and scientists

Theodore Wildi

President, Sperika Enterprises Ltd.

Professor Emeritus of Laval University
Department of Electrical Engineering

SPERIKA ENTERPRISES LTD.
Québec

Cover: Michaud and Associates, graphic designers

Published by Sperika Enterprises Ltd.
P.O. Box 425, Sillery
(Québec) Canada G1T 2R8

15   14   13   12   11   10   9   8   7   6   5   4   3   2   1

Legal deposit: first Quarter 1988

Printed in Canada

PREFACE

As systems of measurement evolved in various parts of the world, they produced a large number of units. This handbook shows in an exceptionally clear and useful way how these units are related to each other, and how they are defined. Its principal advantage lies in the conversion of units, a process made exceedingly simple by a set of new conversion charts. They enable the engineer, scientist and technician to make rapid and clear-cut conversions between units of the American Customary system, the English system, former metric systems, and the International System of Units (SI).

The conversion charts rank the units by order of size so that the relationship between any two units can be found quickly and without ambiguity. They significantly reduce the time usually needed to consult handbooks, tables and so forth, in solving engineering and scientific problems.

The information contained herein is based upon the latest data on units and quantities published by the American National Standards Institute, The Institute of Electrical and Electronics Engineers, the Bureau International des Poids et Mesures, the Committee on Data for Science and Technology (CODATA) and by the International Organization for Standardization.

The conversion methodology was first devised by the author in the early 1970's, and since then thousands of engineers and scientists have discovered its usefulness. The methodology was explained in two pioneering books "Units" and "Understanding Units" which are now out of print, having been replaced by this current up-dated book.

I want to express my thanks to the many individuals who offered their helpful comments. In particular, I want to mention Dr. Gilles Y. Delisle, Dr. Hugh Preston-Thomas, Mr. Guy W.-Richard, Dr. Jacques Vanier and Mr. K. C. Ford.

Last, but not least, I thank Lucie Veilleux for typing the manuscript and putting it into computerized form, and my son Karl for drawing the charts.

T.W.

CONTENTS

INTRODUCING THE CONVERSION CHARTS   1

Methodology   1
Flyers   3
SI Units   3
Symbols   3
Numerical values   3
Converting to SI base units   4
Multiples, submultiples and prefixes   4
Writing numbers   4
Handling special conversions   4

CONVERSION CHARTS

prefixes: symbols and values   8
acceleration   9
amount of substance   10
angle   9
angular velocity   11
angular frequency   11
area   12
concentration   13
density (mass per unit volume)   14
electricity
    capacitance   15
    conductance   15
    current   17
    current density   16
    electric charge   17
    electric field strength   16
    electromotive force   17

electricity (cont'd)

    inductance   15

    resistance   15

    mass resistivity   18

    volume resistivity   18

energy   19

energy (special calorie units)   20

energy (special Btu units)   21

energy (atomic and physical relationships)   22

flow   23

force   24

frequency   24

length   25

length (atomic units and quantities)   26

light

    illuminance   27

    luminance   27

    luminous flux   27

    luminous intensity   27

magnetic moment (atomic units and quantities)   28

magnetism

    magnetic field strength   29

    magnetic flux   29

    magnetic flux density   29

    magnetomotive force   29

mass   30

mass (atomic units and quantities)   31

molality   32

moment of inertia   33

power   34

power density   35

power (special horsepower units)   36

power level difference   36

pressure   37

radiology

    absorbed dose   39

    activity of radionuclides   38

    dose equivalent   38

    exposure   39

speed (see velocity)

stress   37

temperature   40

temperature interval   40

thermal conductivity   41

time   42

torque   43

velocity   44

viscosity

    dynamic viscosity   45

    kinematic viscosity   45

volume (including liquid capacity)   46

volume (including dry capacity)   47

APPENDIX I   SELECTED PHYSICAL CONSTANTS   49

APPENDIX II   THE INTERNATIONAL SYSTEM OF UNITS   50

APPENDIX III   DECIBELS AND NEPERS   54

APPENDIX IV   QUANTITIES AND UNITS   57

APPENDIX V   QUANTITY EQUATIONS AND NUMERICAL EQUATIONS   61

BIBLIOGRAPHY   71

INDEX   73

Introducing The Conversion Charts

Methodology

In this book, the units of various quantities such as force, pressure, viscosity, etc., are displayed by means of conversion charts.

The charts show the relative size of a unit by the position it occupies on the page. The largest unit is at the top, the smallest at the bottom and intermediate units are ranked in between.

The units are connected by arrows, each of which bears a number. This number is the ratio of the larger to the smaller of the units that are connected, and hence its value is always greater than one. The arrow always points toward the smaller of the two units.

In Figure 1, for example, six units of length - the mile, meter, yard, foot, inch and millimeter - are positioned in descending order of size*. The numbers show the relative magnitude of the connected units; the yard is 3 times larger than the foot, the foot is 12 times larger than the inch, and so on. With this arrangement we can convert any UNIT A into any UNIT B by the following simple procedure. Starting from UNIT A we follow any path that ends up at UNIT B while observing the rule:

> with the arrow   MULTIPLY
>
> against the arrow   DIVIDE

Because the arrows point downwards, we multiply when moving down the chart, and divide when moving up.

---

* In most English-speaking countries, including Canada, the SI unit of length is spelled metre, and a unit of volume is spelled litre. On the other hand, the spelling of these two units in the United States is usually meter and liter. In this book, we have adopted the meter, liter spelling.

Example 1:

Convert 2.5 feet to millimeters.

Solution:

Starting from **foot** and moving towards **millimeter** we always move in the direction of the arrows. We must therefore *multiply* the numbers associated with each line:

$$2.5 \text{ feet} = 2.5 \ (\times \ 12) \ (\times \ 25.4) \text{ millimeters}$$
$$= 762 \text{ millimeters}$$

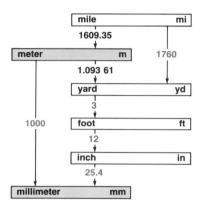

Figure 1

Example 2:

Convert 95 040 inches into miles.

Solution:

Starting from **inch** and moving toward **mile**, we always move against the direction of the arrows. We must therefore *divide* by the number associated with each line:

$$95\ 040 \text{ inches} = 95\ 040 \ (\div \ 12) \ (\div \ 3) \ (\div \ 1760) \text{ miles}$$

$$= \frac{95\ 040}{12 \ \times \ 3 \ \times \ 1760} \text{ miles}$$

$$= 1.5 \text{ miles}$$

Flyers

When units are far apart, several multiplications or divisions must be made to convert from one unit to another. To reduce the arithmetic, "flyers" are introduced to bypass a series of units. A flyer also bears a number, whose value is equal to the product of the numbers which the flyer bypasses. In Figure 1, for example, the flyer 1000 joining meter-millimeter is equal to the product of the numbers $1.093\ 61 \times 3 \times 12 \times 25.4$. ($1.093\ 61$ represents 6-figure accuracy).

Flyers provide additional paths between units, and any of these paths may be followed without affecting the conversion result.

Example 3:
Convert 750 inches into meters.

Solution a:
$$750 \text{ inches } = 750\ (\div\ 12)\ (\div\ 3)\ (\div\ 1.093\ 61)$$
$$= 19.05 \text{ meters}$$

Solution b:
$$750 \text{ inches } = 750\ (\times\ 25.4)\ (\div\ 1000)$$
$$= 19.05 \text{ meters}$$

note that solution b using the flyer is quicker because it involves only one multiplication (with the arrow) and one division (against the arrow).

SI units

SI units and their multiples and submultiples appear in *red* boxes on the conversion charts. Because they are connected by flyers that are multiples of ten, it is possible to move swiftly from one end of a chart to the other should a conversion between widely-separated units be required. American customary units and non-SI metric units can readily be converted to SI or vice-versa by the technique we have just described.

Symbols

The symbols for all SI units (and for many non-SI units) appear on the right-hand side of each box. Symbols have been omitted whenever a clear and authoritative consensus was lacking.

Numerical Values

Numerical values between two units are shown either in red or in black.

RED numbers are exact, by definition.

BLACK numbers are accurate to the number of significant figures shown.

A useful feature of the charts is that exact conversion accuracy between units can be obtained by following paths that involve red numbers. These paths are usually erratic, and may require one or more multiplications and divisions.

For example, referring to the chart on LENGTH, page 25, we find that 1 meter = 39.3701 inches (6-figure accuracy). But if we follow a path involving red numbers we find that 1 meter = 1 ($\times$ 1000) ($\div$ 25.4) = 39.370 078 . . . inches *exactly*.

Converting to SI Base Units

Any derived unit of the SI can be expressed in terms of one or more of the seven SI base units. This "base dimension" is displayed in brackets on the left-hand side of each conversion chart. For example, referring to the chart on ENERGY, page 19, we see that the SI base dimension of the joule is $kg \cdot m^2/s^2$. This relationship is more than one of simple equivalence: 1 joule is *exactly* equal to 1 $kg \cdot m^2/s^2$. Appendix V describes how the dimension of a unit is established.

Multiples and submultiples

Multiples and submultiples of SI units carry prefixes according to Table 1. Using this prefix designation, a length of 1000 meters is called a *kilo*meter, while 1/1000 of a meter (0.001 meter) is called a *milli*meter. Similarly, 1/10 of a liter is called one *deci*liter and 100 liters is one *hecto*liter.

However, in the interest of simplicity, and to minimize the possibilities of error, the prefixes centi, deci, deca and deci are often avoided. For example, builders and machinery manufacturers commonly quote lengths only in meters and millimeters.

Writing Numbers

To simplify the reading of numbers having many digits, the digits are set in groups of three, separated by a space. The spaces are counted to the left and to the right of the decimal marker. This style eliminates the need for commas as group separators. For example, we write

13 754 instead of 13,754

25 146.241 32 instead of 25,146.24132

From an international standpoint, this group-of-three rule also prevents confusion because many countries use the comma as a decimal marker.

Handling special conversions

In solving some engineering problems, the answer may contain several unusual units that seem quite unrelated to the solution we are seeking. Such units should be first expressed in terms of SI base units. The new units can then be simplified to yield the result we are seeking.

Example 4:

In braking a train, it was found that the amount of energy consumed was

$$\text{energy} = 75 \times 10^7 \; \frac{\text{ton} \cdot \text{yard} \cdot \text{mile}}{\text{hour} \cdot \text{minute}}$$

We want to express this unusual result in terms of Btu.

TABLE 1   SI PREFIXES AND THEIR SYMBOLS

Multiplier	Exponent Form	Prefix	SI Symbol
1 000 000 000 000 000 000	10^{18}	exa	E
1 000 000 000 000 000	10^{15}	peta	P
1 000 000 000 000	10^{12}	tera	T
1 000 000 000	10^9	giga	G
1 000 000	10^6	mega	M
1 000	10^3	kilo	k
100	10^2	hecto	h
10	10^1	deca	da
0.1	10^{-1}	deci	d
0.01	10^{-2}	centi	c
0.001	10^{-3}	milli	m
0.000 001	10^{-6}	micro	μ
0.000 000 001	10^{-9}	nano	n
0.000 000 000 001	10^{-12}	pico	p
0.000 000 000 000 001	10^{-15}	femto	f
0.000 000 000 000 000 001	10^{-18}	atto	a

Solution:
Letting {H} be the number of Btu, we write the equation

$$\{H\} \text{ Btu} = 75 \times 10^7 \; \frac{\text{ton} \cdot \text{yard} \cdot \text{mile}}{\text{hour} \cdot \text{minute}}$$

thus

$$\{H\} = 75 \times 10^7 \; \frac{\text{ton} \cdot \text{yard} \cdot \text{mile}}{\text{hour} \cdot \text{minute} \cdot \text{Btu}}$$

Next, using the charts, we convert each unit to SI base units, as follows:

1 ton = 1 (÷ 1.102 31) (× 1000) kg = 907.186 kg   (see p. 30)

1 yard = 1 (÷ 1.093 61) m = 0.914 40 m   (see p. 25)

1 mile = 1 (× 1.609 344) (× 1000) m = 1609.34 m

1 hour = 3600 s

1 minute = 60 s

1 Btu = 1 (× 1.054 35) (× 1000) J = 1 054.35 kg·m^2·s^{-2} (see p. 19)

Note that Btu is converted into SI base units by using the dimension of the joule, displayed in brackets on the left-hand side of the ENERGY chart.
   Substituing these values in the above equation, we obtain

$$\{H\} = 75 \times 10^7 \; \frac{(907.186 \text{ kg}) \times (0.9144 \text{ m}) \times (1609.34 \text{ m})}{(3600 \text{ s}) \times (60 \text{ s}) \times (1054.35 \text{ kg·m}^2\text{·s}^{-2})}$$

$$= 4.396 \times 10^6$$

Note that all units cancel, showing that the original answer did indeed have the dimension of energy. The energy consumed is therefore 4.396×10^6 Btu.

Conversion Charts

SI UNITS:	RED BOXES	

ALL OTHER UNITS:	WHITE BOXES	

RED NUMBERS are exact, by definition

BLACK NUMBERS are accurate to the number
of significant figures shown

Conversion Rule

WITH THE ARROW - MULTIPLY

AGAINST THE ARROW - DIVIDE

SI PREFIXES
SYMBOLS AND VALUES

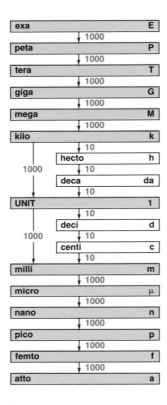

ACCELERATION

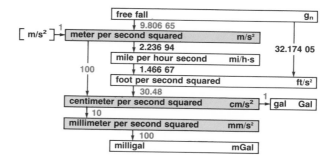

ANGLE

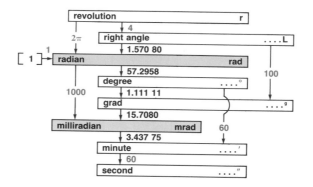

AMOUNT OF SUBSTANCE

10

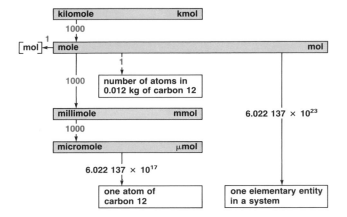

ANGULAR VELOCITY

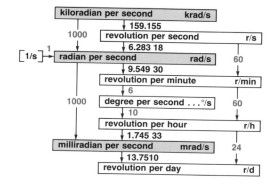

Example: Convert 1800 r/min to radians per second

Solution: 1800 r/min = 1800 (÷ 9.549 30) rad/s = 188.495 rad/s

ANGULAR FREQUENCY
CIRCULAR FREQUENCY
PULSATANCE

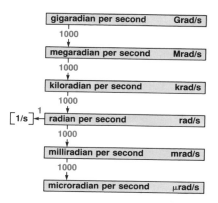

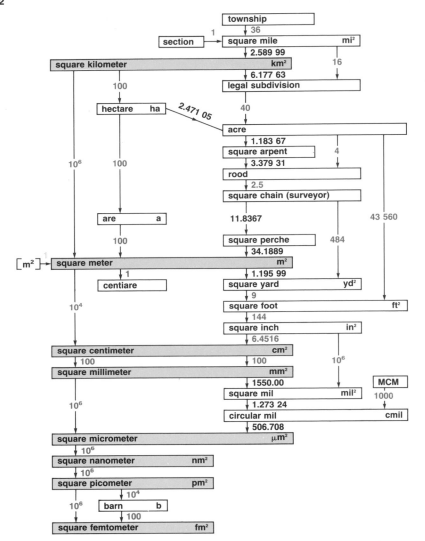

Example: Convert 4 square miles to hectares

Solution: 4 mi² = 4 (× 2.589 99) (× 100) ha = 1036 hectares

CONCENTRATION

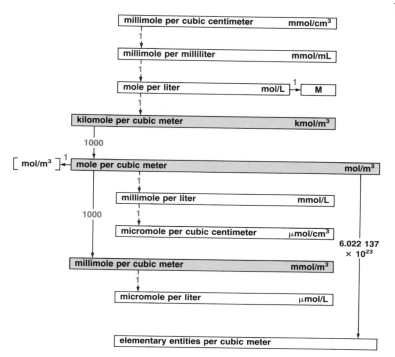

millimole per cubic centimeter	mmol/cm³
millimole per milliliter	mmol/mL
mole per liter	mol/L	M
kilomole per cubic meter	kmol/m³

1000

[mol/m³] 1 — mole per cubic meter   mol/m³

| millimole per liter | mmol/L |
| micromole per cubic centimeter | μmol/cm³ |

1000

6.022 137 × 10²³

| millimole per cubic meter | mmol/m³ |
| micromole per liter | μmol/L |

| elementary entities per cubic meter | |

DENSITY (mass per unit volume)

14

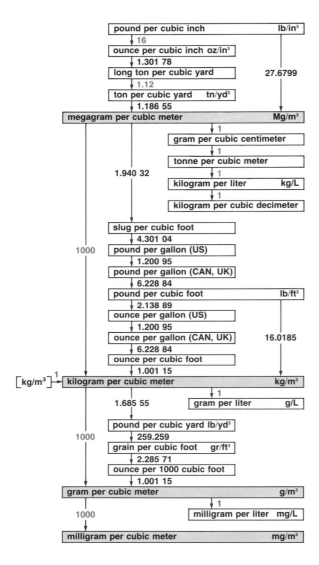

pound per cubic inch	lb/in³
↓ 16	
ounce per cubic inch oz/in³	
↓ 1.301 78	
long ton per cubic yard	27.6799
↓ 1.12	
ton per cubic yard tn/yd³	
↓ 1.186 55	
megagram per cubic meter	**Mg/m³**

↓ 1
gram per cubic centimeter
↓ 1
tonne per cubic meter
↓ 1
kilogram per liter   kg/L
↓ 1
kilogram per cubic decimeter

1.940 32

slug per cubic foot
↓ 4.301 04
pound per gallon (US)
↓ 1.200 95
pound per gallon (CAN, UK)
↓ 6.228 84
pound per cubic foot   lb/ft³
↓ 2.138 89
ounce per gallon (US)
↓ 1.200 95
ounce per gallon (CAN, UK)   16.0185
↓ 6.228 84
ounce per cubic foot
↓ 1.001 15

1000

[kg/m³] → **kilogram per cubic meter**   **kg/m³**

1.685 55   ↓ 1   gram per liter   g/L

pound per cubic yard lb/yd³
↓ 259.259
grain per cubic foot   gr/ft³
↓ 2.285 71
ounce per 1000 cubic foot
↓ 1.001 15

1000

gram per cubic meter   **g/m³**

↓ 1   milligram per liter   mg/L

1000

milligram per cubic meter   **mg/m³**

Example: Convert 60 lb per cubic foot to tons per cubic yard
Solution: 60 lb/ft³ = 60 (× 16.0185) (÷ 1000) (÷ 1.186 55) = 0.81 ton/yd³

ELECTRICITY

RESISTANCE

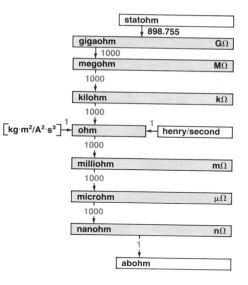

statohm	
↓ 898.755	
gigaohm	GΩ
↓ 1000	
megohm	MΩ
↓ 1000	
kilohm	kΩ
↓ 1000	
[kg·m²/A²·s³] →¹ ohm ←¹ henry/second	
↓ 1000	
milliohm	mΩ
↓ 1000	
microhm	μΩ
↓ 1000	
nanohm	nΩ
↓ 1	
abohm	

CONDUCTANCE

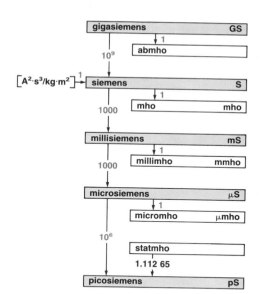

gigasiemens	GS
↓ 1	
abmho	
10⁹	
[A²·s³/kg·m²] →¹ siemens	S
↓ 1	
mho	mho
1000	
millisiemens	mS
↓ 1	
millimho	mmho
1000	
microsiemens	μS
↓ 1	
micromho	μmho
10⁶	
statmho	
1.112 65	
picosiemens	pS

INDUCTANCE

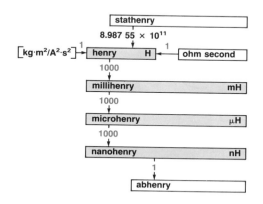

stathenry	
8.987 55 × 10¹¹	
[kg·m²/A²·s²] →¹ henry H ←¹ ohm second	
↓ 1000	
millihenry	mH
↓ 1000	
microhenry	μH
↓ 1000	
nanohenry	nH
↓ 1	
abhenry	

CAPACITANCE

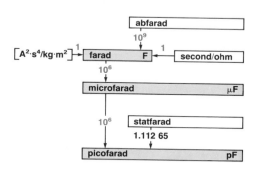

abfarad	
10⁹	
[A²·s⁴/kg·m²] →¹ farad F ←¹ second/ohm	
10⁶	
microfarad	μF
10⁶	statfarad
	1.112 65
picofarad	pF

ELECTRICITY

CURRENT DENSITY

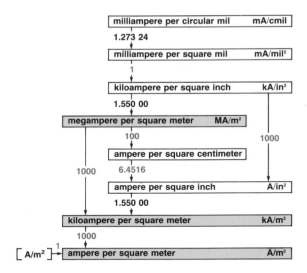

ELECTRIC FIELD STRENGTH

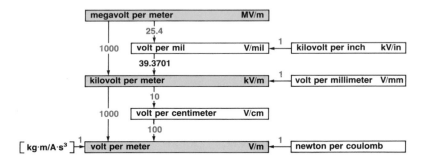

ELECTRICITY

CURRENT

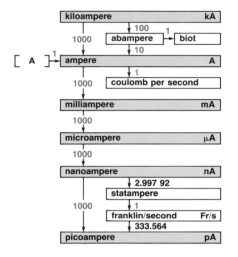

CHARGE

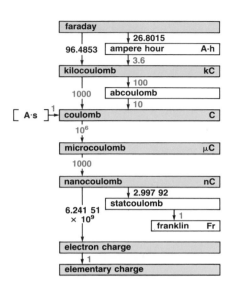

ELECTROMOTIVE FORCE

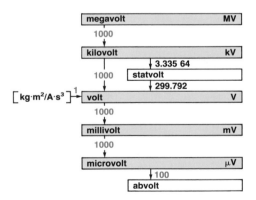

Example: Express 3 kilovolts in terms of SI base dimensions

Solution: $3 \text{ kV} = 3 \ (\times \ 1000) \ (\times \ 1) \ \text{kg·m}^2/(\text{A·s}^3) = 3000 \ \text{kg·m}^2/(\text{A·s}^3)$

ELECTRICITY

MASS RESISTIVITY

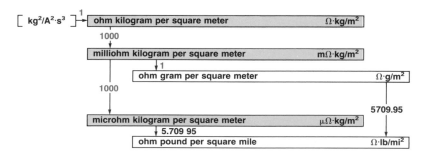

VOLUME RESISTIVITY

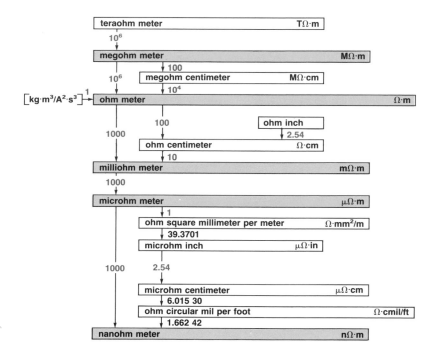

Example: Convert 60 ohm circular mil per foot to ohm cm

Solution: 60 Ωcmil/ft = 60 ($\times$ 1.662 42) ($\div$ 1000) ($\div$ 1000) ($\div$ 10) = 9.974 52 $\times$ 10^{-6} Ω·cm

ENERGY

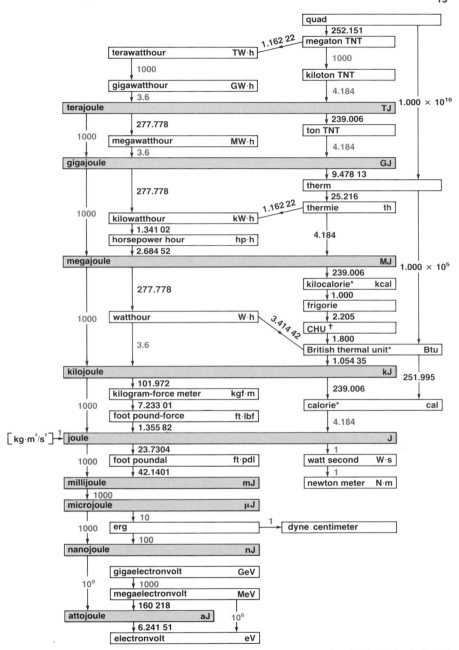

Note: Btu and calorie are *thermochemical* units

CHU is the heat required to raise the temperature of 1 lb of water by 1°C

ENERGY (calorie units)

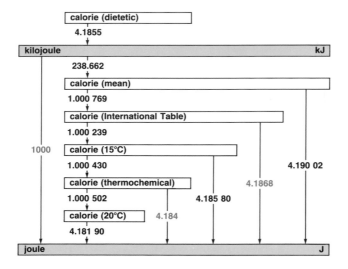

ENERGY (Btu units)

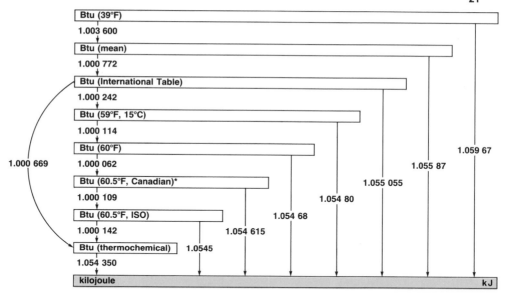

Btu (39°F)
1.003 600

Btu (mean)
1.000 772

Btu (International Table)
1.000 242

Btu (59°F, 15°C)
1.000 114

Btu (60°F)
1.000 062

Btu (60.5°F, Canadian)*
1.000 109

Btu (60.5°F, ISO)
1.000 142

Btu (thermochemical)   1.0545
1.054 350

1.000 669

1.054 615

1.054 68

1.054 80

1.055 055

1.055 87

1.059 67

kilojoule   kJ

Example: Convert 4000 Btu(60°F) to Btu (thermochemical)
Solution: 4000 Btu(60°F) = 4000 (× 1.054 68) (÷ 1.054 350) Btu(thermochemical) = 4001.25 Btu(thermochemical)

ENERGY (atomic and physical relationships)

c = velocity of light = 2.997 924 58 × 10⁸ m/s = {c} m/s

k = Boltzmann constant = 1.380 658 × 10⁻²³ J/K = {k} J/K

h = Planck constant = 6.626 0755 × 10⁻³⁴ J·s = {h} J·s

c₂ = second radiation constant = 1.438 769 × 10⁻²m·K = {c₂} m·K

N_A = Avogadro constant = 6.022 1367 × 10²³ mol⁻¹ = {N_A} mol⁻¹

{R_∞} = Rydberg constant = 1.097 3731 × 10⁷ m⁻¹ = {R_∞} m⁻¹

{ } = numerical value of constant when its dimension is expressed in SI units

cal_th = thermochemical calorie

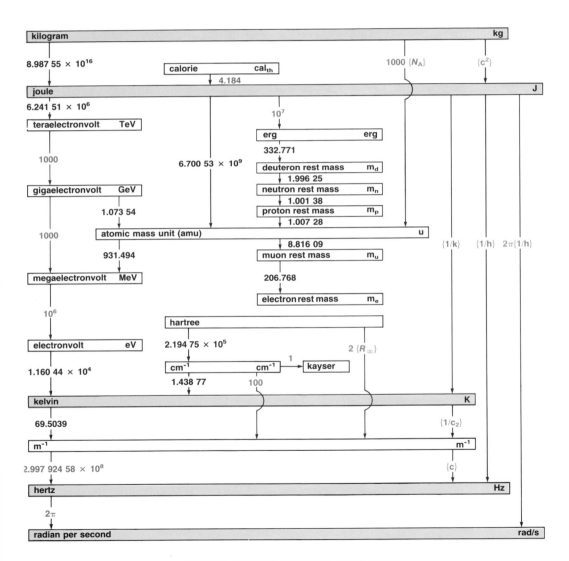

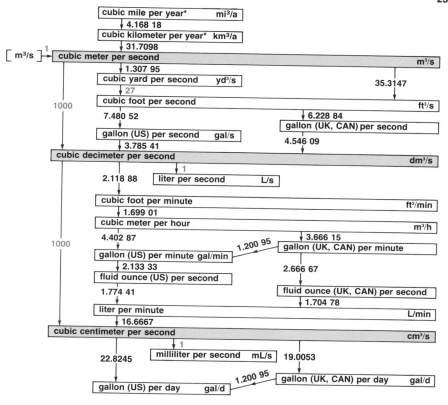

cubic mile per year* mi³/a
↓ 4.168 18
cubic kilometer per year* km³/a
↓ 31.7098
[m³/s]→ 1 cubic meter per second m³/s
↓ 1.307 95
cubic yard per second yd³/s
1000 ↓ 27
cubic foot per second ft³/s
7.480 52 6.228 84
↓ ↓
gallon (US) per second gal/s gallon (UK, CAN) per second
↓ 3.785 41 4.546 09
cubic decimeter per second dm³/s
↓ 1
2.118 88 liter per second L/s
cubic foot per minute ft³/min
↓ 1.699 01
cubic meter per hour m³/h
4.402 87 3.666 15
1000 1.200 95 ↓
gallon (US) per minute gal/min gallon (UK, CAN) per minute
↓ 2.133 33 2.666 67
fluid ounce (US) per second
1.774 41 fluid ounce (UK, CAN) per second
↓ ↓ 1.704 78
liter per minute L/min
↓ 16.6667
cubic centimeter per second cm³/s
↓ 1
22.8245 milliliter per second mL/s 19.0053
↓ 1.200 95 ↓
gallon (US) per day gal/d gallon (UK, CAN) per day gal/d

*year = 365 days.

Example: Convert 250 gallons(US) per minute to liters per second

Solution: 250 gal(U.S.)/min = 250 (× 2.13333) (× 1.774 41) (× 16.6667) (÷ 1000) L/s = 15.7725 L/s

FORCE

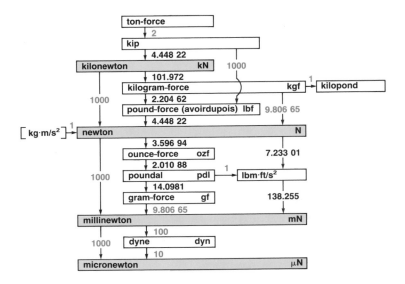

Example: Convert 30 lbf to newtons

Solution: 30 lbf = 30 (× 4.448 22) N = 133.447 N

FREQUENCY

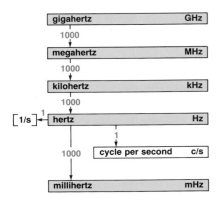

LENGTH

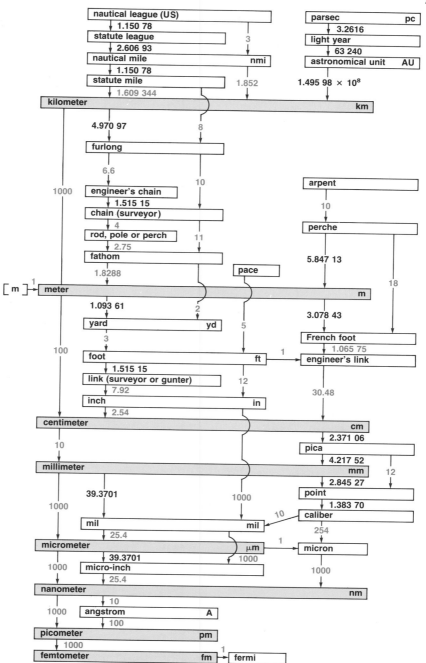

LENGTH (atomic units and quantities)

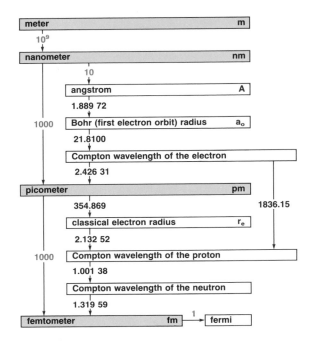

Example: Express the Bohr first electron radius (a_o) in fermis

Solution: a_o = 1 (× 21.8100) (× 2.426 31) (× 1000) fermi = 52 917.8 fermi

LIGHT

LUMINANCE

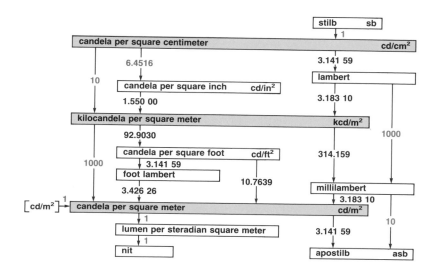

ILLUMINANCE

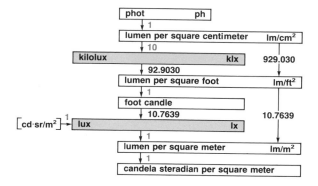

LUMINOUS INTENSITY

LUMINOUS FLUX

MAGNETIC MOMENT (atomic units and quantities)

28

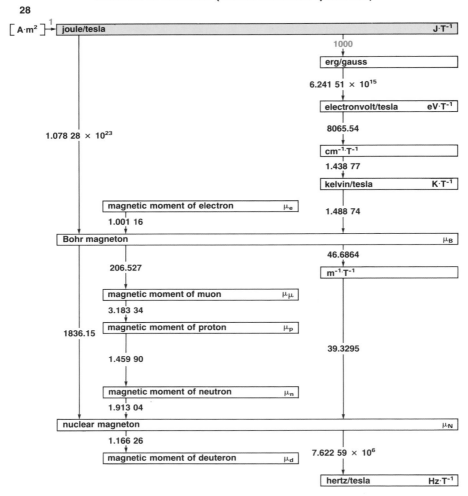

$[A \cdot m^2]$ — joule/tesla — $J \cdot T^{-1}$

1000

erg/gauss

$6.241\ 51 \times 10^{15}$

electronvolt/tesla — $eV \cdot T^{-1}$

8065.54

$cm^{-1} \cdot T^{-1}$

1.438 77

kelvin/tesla — $K \cdot T^{-1}$

$1.078\ 28 \times 10^{23}$

magnetic moment of electron — μ_e

1.001 16

1.488 74

Bohr magneton — μ_B

46.6864

$m^{-1} \cdot T^{-1}$

206.527

magnetic moment of muon — μ_μ

3.183 34

magnetic moment of proton — μ_p

1836.15

1.459 90

39.3295

magnetic moment of neutron — μ_n

1.913 04

nuclear magneton — μ_N

1.166 26

7.622 59 × 10^6

magnetic moment of deuteron — μ_d

hertz/tesla — $Hz \cdot T^{-1}$

MAGNETISM

MAGNETOMOTIVE FORCE

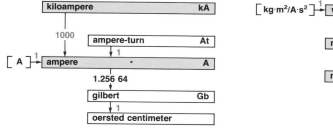

kiloampere	kA

1000

ampere-turn	At

1

[A] → ampere	· A

1.256 64

gilbert	Gb

1

oersted centimeter	

MAGNETIC FLUX

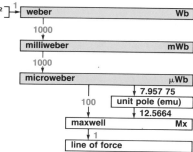

[kg·m²/A·s²] → weber	Wb

1000

milliweber	mWb

1000

microweber	μWb

7.957 75

unit pole (emu)	

100   12.5664

maxwell	Mx

1

line of force	

MAGNETIC FIELD STRENGTH

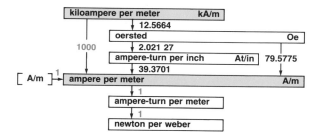

kiloampere per meter	kA/m

12.5664

oersted	Oe

2.021 27

ampere-turn per inch	At/in	79.5775

1000   39.3701

[A/m] → ampere per meter	A/m

1

ampere-turn per meter	

1

newton per weber	

MAGNETIC FLUX DENSITY

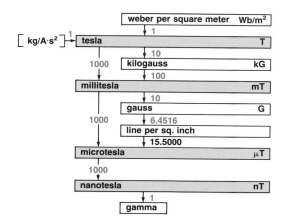

weber per square meter	Wb/m²

1

[kg/A·s²] → tesla	T

10

1000 kilogauss	kG

100

millitesla	mT

10

gauss	G

1000   6.4516

line per sq. inch	

15.5000

microtesla	μT

1000

nanotesla	nT

1

gamma	

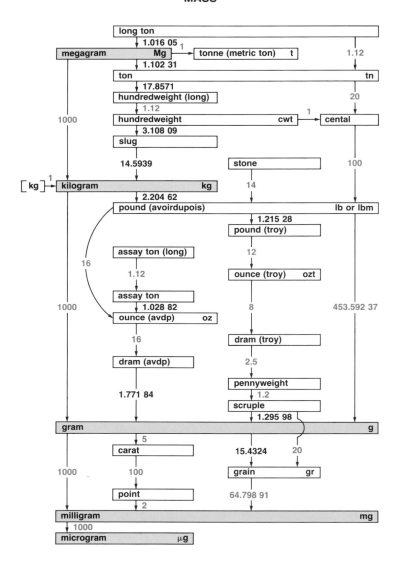

Example: Convert 2 kg to troy ounces **exactly**

Solution: For exact conversions we follow the arrows bearing red numbers. Thus:

$$2\,kg = 2(\times\,1000)\,(\times\,1000)\,(\div\,64.798\,91)\,(\div\,20)\,(\div\,1.2)\,(\div\,2.5)\,(\div\,8)\,ozt = 64.301\,49\ .\ .\ .\ ..\ ozt$$

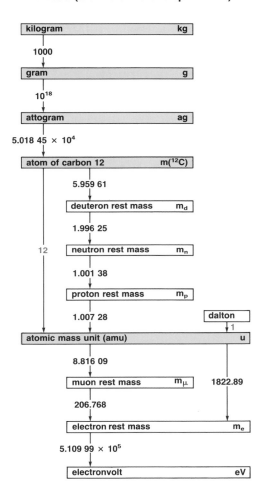

MOLALITY

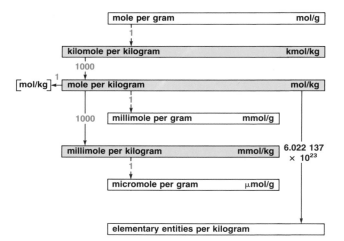

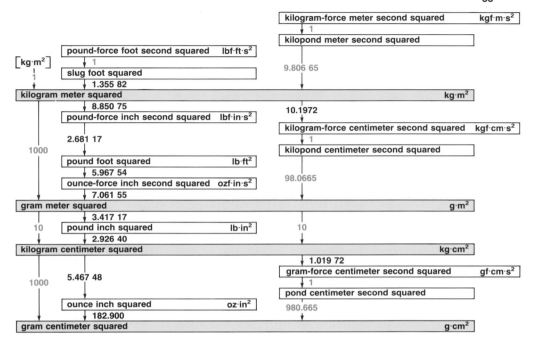

Example: Convert 80 lb·ft^2 into kg·m^2

Solution: 80 lb·ft^2 = 80 (× 5.967 54) (× 7.061 55) (÷ 1000) kg·m^2 = 3.371 21 kg·m^2

POWER

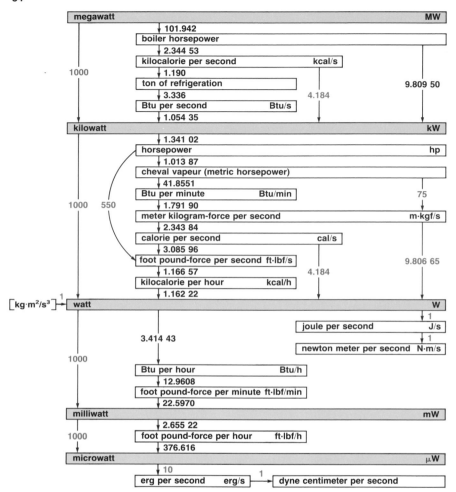

megawatt		MW
↓ 101.942		
boiler horsepower		
↓ 2.344 53		
kilocalorie per second	kcal/s	
↓ 1.190		
ton of refrigeration		9.809 50
↓ 3.336	4.184	
Btu per second	Btu/s	
↓ 1.054 35		

kilowatt		kW
↓ 1.341 02		
horsepower		hp
↓ 1.013 87		
cheval vapeur (metric horsepower)		
↓ 41.8551		75
Btu per minute	Btu/min	
↓ 1.791 90		
meter kilogram-force per second		m·kgf/s
↓ 2.343 84		
calorie per second	cal/s	
↓ 3.085 96		9.806 65
foot pound-force per second ft·lbf/s		
↓ 1.166 57	4.184	
kilocalorie per hour	kcal/h	
↓ 1.162 22		

1000   550

$[kg·m^2/s^3]$ — 1

watt		W
		↓ 1
		joule per second J/s
3.414 43		↓ 1
		newton meter per second N·m/s
Btu per hour	Btu/h	
↓ 12.9608		
foot pound-force per minute ft·lbf/min		
↓ 22.5970		

1000

milliwatt		mW
↓ 2.655 22		
foot pound-force per hour	ft·lbf/h	
↓ 376.616		

1000

microwatt		μW
↓ 10		
erg per second erg/s	→ 1	**dyne centimeter per second**

Note: Btu and calorie are *thermochemical* units

Example: Convert 12 boiler horsepower into kilowatts
Solution: 12 boiler hp = 12 (÷ 101.942) (× 1000) kW = 117.714 kW

POWER DENSITY

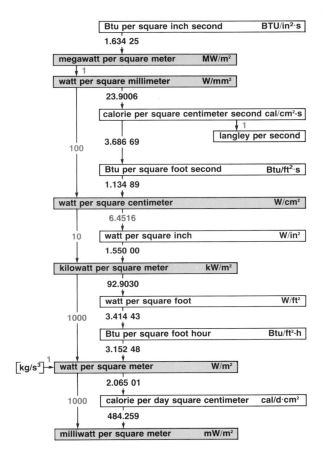

| Btu per square inch second | BTU/in²·s |

1.634 25

megawatt per square meter   MW/m²

1

watt per square millimeter   W/mm²

23.9006

calorie per square centimeter second cal/cm²·s

1

langley per second

3.686 69

Btu per square foot second   Btu/ft²·s

1.134 89

watt per square centimeter   W/cm²

6.4516

10   watt per square inch   W/in²

1.550 00

kilowatt per square meter   kW/m²

92.9030

watt per square foot   W/ft²

1000   3.414 43

Btu per square foot hour   Btu/ft²·h

3.152 48

100

1000

[kg/s³] → 1 → watt per square meter   W/m²

2.065 01

1000   calorie per day square centimeter   cal/d·cm²

484.259

milliwatt per square meter   mW/m²

Note: Btu and calorie are *thermochemical* units

POWER (horsepower units)

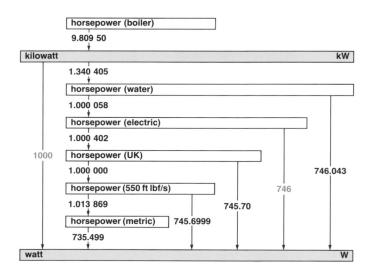

POWER LEVEL DIFFERENCE

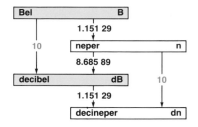

Example: Convert 80 decibels to nepers
Solution: 80 dB = 80 (÷ 8.685 89)Np = 9.21 Np

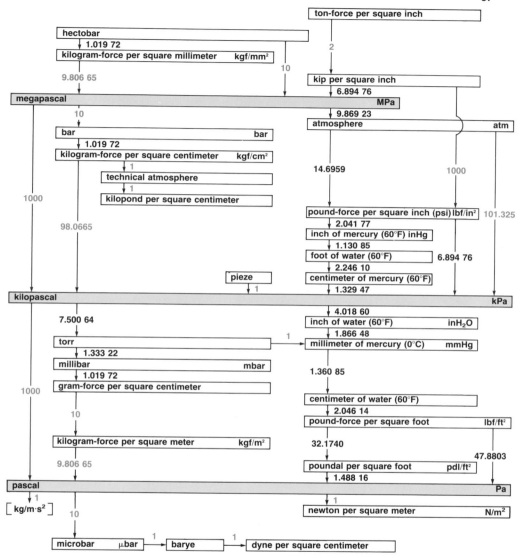

ton-force per square inch

hectobar
↓ 1.019 72
kilogram-force per square millimeter   kgf/mm²

9.806 65

2

10

kip per square inch
↓ 6.894 76

megapascal
MPa

10

↓ 9.869 23
atmosphere   atm

bar   bar
↓ 1.019 72
kilogram-force per square centimeter   kgf/cm²
↓ 1
technical atmosphere
↓ 1
kilopond per square centimeter

14.6959

1000

1000

98.0665

pound-force per square inch (psi) lbf/in²   101.325
↓ 2.041 77
inch of mercury (60°F) inHg
↓ 1.130 85
foot of water (60°F)   6.894 76
↓ 2.246 10

pieze
↓ 1

centimeter of mercury (60°F)
↓ 1.329 47

kilopascal
kPa

7.500 64

↓ 4.018 60
inch of water (60°F)   inH₂O
↓ 1.866 48

torr
↓ 1.333 22

1

millimeter of mercury (0°C)   mmHg

millibar   mbar
↓ 1.019 72
gram-force per square centimeter

1.360 85

1000

centimeter of water (60°F)
↓ 2.046 14
pound-force per square foot   lbf/ft²

10

kilogram-force per square meter   kgf/m²

32.1740

47.8803

9.806 65

poundal per square foot   pdl/ft²
↓ 1.488 16

pascal
Pa

$\left[\text{kg/m·s}^2 \right]$ ↓ 1

↓ 1

10

newton per square meter   N/m²

microbar   μbar → barye → dyne per square centimeter

1   1

Example: Convert 15 inches of water(60°F) to kilopascals

Solution: 15 inH₂0(60°F) = 15 (÷ 4.018 60) kPa = 3.732 64 kPa

RADIOLOGY

DOSE EQUIVALENT

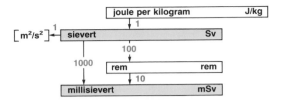

ACTIVITY (of radionuclides)

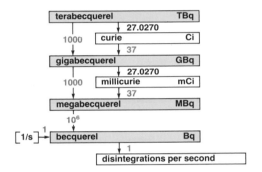

Example: Convert 5 millicuries to gigabecquerels

Solution: 5 mCi = 5 (÷ 27.0270) GBq = 0.185 GBq

RADIOLOGY

EXPOSURE

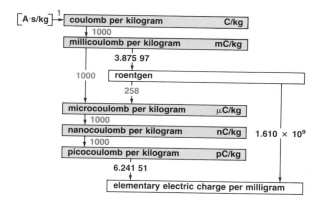

ABSORBED DOSE

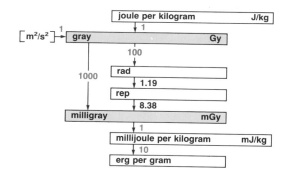

Example: Express 200 rep in terms of joules per kilogram
Solution: 200 rep = 200 (× 8.38) (÷ 1000) (÷ 1) J/kg = 1.68 J/kg

TEMPERATURE

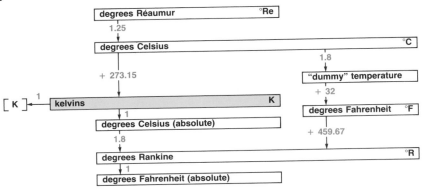

Note: Celsius was formerly called centigrade

The «dummy» temperature is simply a convenient interface between °F and °C

Examples of temperature conversion:

80 degrees Réaumur = (80 × 1.25)°C = 100°C

100 degrees Celsius = (100 + 273.15) K = 373.15 K

100 degrees Celsius = (100 × 1.8) + 32°F = 212°F

108 degrees Fahrenheit = (108 + 459.67) ÷ 1.8 K = (567.67 ÷ 1.8) K = 315.372 K

TEMPERATURE INTERVAL

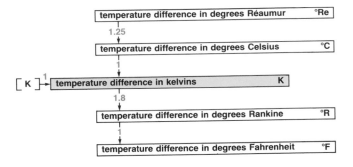

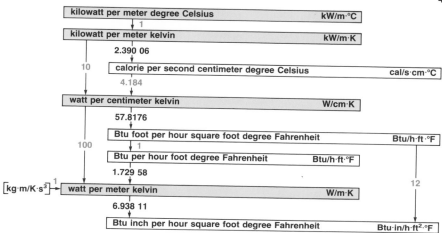

Note: Btu and calorie are *thermochemical* units

Example: Convert 500 Btu/(h·ft·°F) into watts per meter kelvin
Solution: 500 Btu/(h·ft·°F) = 500 (× 1.729 58) W/(m·K) = 864.79 W/(m·K)

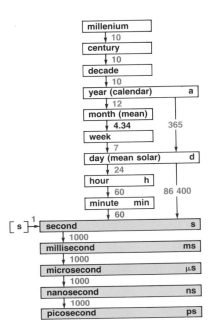

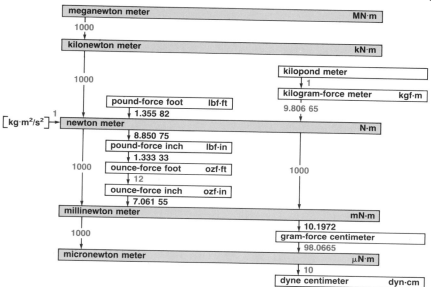

Example: Convert 12000 lbf·ft to kilonewton meters

Solution: 12000 lbf·ft = 12000 (× 1.355 82) (÷ 1000) kN·m = 16.2698 kN·m

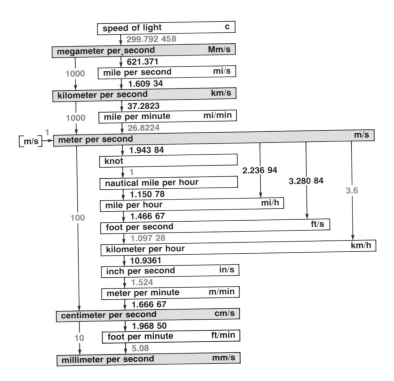

Example: Convert 75 m/s into miles per hour

Solution: 75 m/s = 75 (× 2.236 94) mi/h = 167.77 mi/h

VISCOSITY

DYNAMIC VISCOSITY

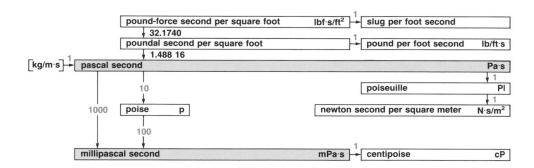

KINEMATIC VISCOSITY

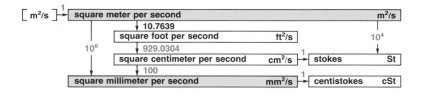

VOLUME (including liquid capacity)

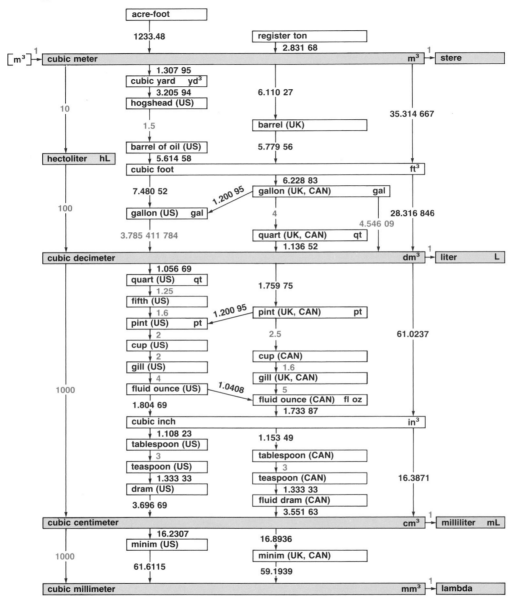

VOLUME (including dry capacity)

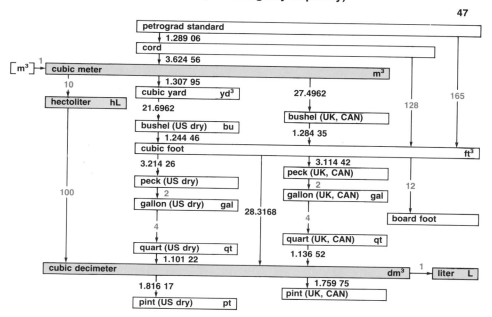

Example: Convert 5000 cubic feet to cubic meters

Solution: 5000 ft³ = 5000 (÷ 128) (× 3.624 56)m³ = 141.584 m³

APPENDIX I

SELECTED PHYSICAL CONSTANTS [**]

Quantity	Quoted Value	S[*]	SI unit	Symbol
Atomic mass constant	$1.660\ 5402 \times 10^{-27}$	10	kg	m_u
Avogadro constant	$6.022\ 1367 \times 10^{23}$	36	mol^{-1}	N_A
Bohr magneton	$9.274\ 0154 \times 10^{-24}$	31	$J \cdot T^{-1}$	μ_B
Boltzmann constant	$1.380\ 658 \times 10^{-23}$	12	$J \cdot K^{-1}$	$k (= R/N_A)$
Electron charge	$1.602\ 177\ 33 \times 10^{-19}$	49	C	$-e$
Electron specific charge	$-1.758\ 819\ 62 \times 10^{11}$	53	$C \cdot kg^{-1}$	$-e/m_e$
Electron rest mass	$9.109\ 3897 \times 10^{-31}$	54	kg	m_e
Faraday constant	$9.648\ 5309 \times 10^4$	29	$C \cdot mol^{-1}$	F
Fine-structure constant	$0.007\ 297\ 353\ 08$	33	—	α
Josephson frequency to voltage ratio	$4.835\ 9767 \times 10^{14}$	14	$Hz \cdot V^{-1}$	$2e/h$
Magnetic flux quantum	$2.067\ 834\ 61 \times 10^{-15}$	61	Wb	ϕ_o
Molar gas constant	$8.314\ 510$	70	$J \cdot mol^{-1} \cdot K^{-1}$	R
Newtonian gravitational constant	$6.672\ 59 \times 10^{-11}$	85	$m^3 \cdot kg^{-1} \cdot s^{-2}$	G
Permeability of vacuum	$4\pi \times 10^{-7}$	d	H/m	μ_o
Permittivity of vacuum	$8.854\ 1878.. \times 10^{-12}$	d	F/m	ϵ_o
Planck constant	$6.626\ 0755 \times 10^{-34}$	40	$J \cdot s$	h
Planck constant/2π	$1.054\ 572\ 66 \times 10^{-34}$	63	$J \cdot s$	$h (= h/2\pi)$
Quantum of circulation	$3.636\ 948\ 07 \times 10^{-4}$	33	$J \cdot s \cdot kg^{-1}$	$h/2m_e$
Rydberg constant	$1.097\ 373\ 1534 \times 10^7$	13	m^{-1}	R_∞
Standard volume of ideal gas	$22.414\ 10 \times 10^{-3}$	19	$m^3 \cdot mol^{-1}$	V_m
Stefan-Boltzmann constant	$5.670\ 51 \times 10^{-8}$	19	$W \cdot K^{-4} \cdot m^{-2}$	σ
Speed of light	$299\ 792\ 458$	d	$m \cdot s^{-1}$	c

[*] The numbers in this column are the one-standard-deviation uncertainties in the last digits of the quoted value. The symbol d stands here for defined value.

[**] Drawn from E.R. Cohen and B.N. Taylor "The 1986 Adjustment of the Fundamental Physical Constants", Committee on Data for Science and Technology (CODATA) Bulletin Number 63, November 1986.

APPENDIX II

THE INTERNATIONAL SYSTEM OF UNITS

In 1960, at the Eleventh General Conference of Weights and Measures held at Sèvres, France, a world system of units was established, officially named "Système international d'unités". The International System of Units is universally designated by the abbreviation SI.

Base and Derived Units of the SI

The foundation of the International System of Units rests upon seven base units, listed in Table A1.

From these base units other units are derived to express quantities such as area, power, force, magnetic flux, and so forth. Some of these derived units occur so frequently that they have been given special names. A few of these special names are listed in Table A2.

Definitions of SI Base Units

The **meter** (m) is the length of the path travelled by light in vacuum during a time interval of 1/299 792 458 of a second.

The **kilogram** (kg) is the unit of mass; it is equal to the mass of the international prototype of the kilogram. *(The international prototype, made of platinum-iridium is preserved in a vault at Sèvres, France, by the International Bureau of Weights and Measures under conditions specified by the 1st CGPM in 1889).*

The **second** (s) is the duration of 9 192 631 770 periods of the radiation corresponding to the transition between the two hyperfine levels of the ground state of the cesium-133 atom.

TABLE A1	BASE UNITS OF THE SI	
Quantity	**Unit**	**Symbol**
Base Units		
Length	meter	m
Mass	kilogram	kg
Time	second	s
Electric current	ampere	A
Temperature	kelvin	K
Luminous intensity	candela	cd
Amount of substance	mole	mol

The **ampere** (A) is that constant current which, if maintained in two straight paral-lel conductors of infinite length, of negligible circular cross section, and placed 1 meter apart in a vacuum, would produce between these conductors a force equal to 2×10^{-7} newton per meter of length.

**TABLE A2   DERIVED UNITS AND SUPPLEMENTARY UNITS
WITH SPECIAL NAMES**

Quantity	Unit	Symbol	Formula
Absorbed dose	gray	Gy	J/kg
Activity (radionuclide)	becquerel	Bq	s^{-1}
Capacitance	farad	F	C/V
Celsius temperature	degree Celsius	°C	K
Dose equivalent	sievert	Sv	J/kg
Electric charge	coulomb	C	A·s
Electric conductance	siemens	S	$1/\Omega$
Electric potential	volt	V	W/A
Electric resistance	ohm	Ω	V/A
Energy	joule	J	N·m
Force	newton	N	$kg \cdot m/s^2$
Frequency	hertz	Hz	1/s
Illumination	lux	lx	$cd \cdot sr/m^2$
Inductance	henry	H	Wb/A
Luminous flux	lumen	lm	cd·sr
Magnetic flux	weber	Wb	V·s
Magnetic flux density	tesla	T	Wb/m^2
Power	watt	W	J/s
Pressure	pascal	Pa	N/m^2
Supplementary Units*			
Plane angle	radian	rad	1
Solid angle	steradian	sr	1

*   In 1980 the Comité International des Poids et Mesures (CIPM) decided to interpret the class of supplementary units in the International System as a class of dimensionless derived units.

The **kelvin** (K), unit of thermodynamic temperature, is the fraction 1/273.16 of the thermodynamic temperature of the triple point of water.

The **candela** (cd) is the luminous intensity, in a given direction, of a source that emits monochromatic radiation of frequency 540×10^{12} hertz and that has a radiant intensity in that direction of 1/683 watt per steradian.

The **mole** (mol) is the amount of substance of a system which contains as many elementary entities as there are atoms in 0.012 kilogram of carbon 12. *Note:* When the mole is used, the elementary entities must be specified and may be atoms, molecules, ions, electrons, other particles or specified groups of such particles.

Definitions of Supplementary Units

The **radian** (rad) is the angle between two radii of a circle which cut off on the circumference an arc equal in length to the radius.

The **steradian** (sr) is the solid angle which, having its vertex in the center of a sphere, cuts off an area of the surface of the sphere equal to that of a square with sides of length equal to the radius of the sphere.

Definitions of Derived Units

Some of the more important derived units are defined as follows:

The **becquerel** (Bq) is the activity of a radionuclide decaying at the rate of one spontaneous nuclear transition per second.

The **coulomb** (C) is the quantity of electricity transported in 1 second by a current of 1 ampere. *(Hence 1 coulomb = 1 ampere second)*

The **degree Celsius** (°C) is equal to the kelvin and is used in place of the kelvin for expressing Celsius temperature (symbol t) defined by the equation $t = T - T_o$ where T is the thermodynamic temperature and $T_o = 273.15$ K, by definition.

The **farad** (F) is the capacitance of a capacitor between the plates of which there appears a difference of potential of 1 volt when it is charged by a quantity of electricity equal to 1 coulomb. *(Hence 1 farad = 1 coulomb per volt)*

The **gray** (Gy) is the absorbed dose when the energy per unit mass imparted to matter by ionizing radiation is one joule per kilogram. *Note:* The gray is also used for the ionizing radiation quantities: specific energy imparted, kerma, and absorbed dose index.

The **henry** (H) is the inductance of a closed circuit in which an electromotive force of 1 volt is produced when the electric current in the circuit varies uniformly at a rate of 1 ampere per second. *(Hence 1 henry = 1 volt second per ampere)*

The **hertz** (Hz) is the frequency of a periodic phenomenon of which the period is 1 second*.

The **joule** (J) is the work done when the point of application of 1 newton is displaced a distance of 1 meter in the direction of the force. *(Hence 1 joule = 1 newton meter)*

The **lumen** (lm) is the luminous flux emitted in a solid angle of 1 steradian by a uniform point source having an intensity of 1 candela. *(Hence 1 lumen = 1 candela steradian)*

The **lux** (lx) is the unit of illuminance equal to one lumen per square meter.

The **newton** (N) is that force which gives to a mass of 1 kilogram an acceleration of 1 meter per second per second. *(Hence 1 newton = 1 kilogram meter per second squared)*

The **ohm** (Ω) is the electric resistance between two points of a conductor when a constant difference of potential of 1 volt, applied between these two points, produces in this conductor a current of 1 ampere, this conductor not being the source of any electromotive force. *(Hence 1 ohm = 1 volt per ampere)*

The **pascal** (P) is the unit of pressure or stress equal to one newton per square meter.

The **siemens** (S) is the unit of electric conductance equal to one reciprocal ohm. *(The siemens was formerly named the mho.)*

The **sievert** (Sv) is the dose equivalent when the absorbed dose of ionizing radiation multiplied by the dimensionless factors Q (quality factor) and N (product of any other multiplying factors) stipulated by the International Commission on Radiological Protection is one joule per kilogram.

The **tesla** (T) is the unit of magnetic flux density equal to one weber per square meter.

The **volt** (V) is the difference of electric potential between two points of a conducting wire carrying a constant current of 1 ampere, when the power dissipated between these points is equal to 1 watt. *(Hence 1 volt = 1 watt per ampere)*

The **watt** (W) is the power which gives rise to the production of energy at the rate of 1 joule per second. *(Hence 1 watt = 1 joule per second)*

The **weber** (Wb) is the magnetic flux which, linking a circuit of one turn, produces in it an electromotive force of 1 volt as it is reduced to zero at a uniform rate in 1 second. *(Hence 1 weber = 1 volt second)*

---

* The period is the duration of one cycle of a periodic phenomenon. One cycle is any portion of a periodic phenomenon that contains the minimum information needed to describe the periodic phenomenon.

APPENDIX III

DECIBELS AND NEPERS

Power ratio and amplitude ratio

The power ratio of a given power P and an arbitrary reference power P_o is given by

$$\text{power ratio} = P/P_o$$

where P_o, by definition, is the denominator.

The amplitude ratio of a given amplitude F and an arbitrary reference amplitude F_o is given by

$$\text{amplitude ratio} = F/F_o$$

where F_o, by definition, is in the denominator.

In electrotechnology, the amplitude is usually a voltage or current, while in acoustics it is usually a pressure.

Power ratios and amplitude ratios are clearly dimensionless numbers.

Power level difference

Power level difference (symbol L_P) is a quantity whose magnitude is related logarithmically to the power ratio. The unit of power level difference is the decibel (symbol dB). The magnitude L_P of the power level difference is given by the equation:

$$L_P = \{10 \log (P/P_o)\}\,[\text{decibel}] \tag{1}$$

In this equation the term $\{10 \log (P/P_o)\}$ is the numerical value of the quantity, while [decibel] is the unit.

For example, if the given power P is 12.6 W and the reference power P_o is 1 mW, the power level difference is

$$
\begin{aligned}
L_P &= 10 \log (P/P_o) \text{ dB} \\
&= 10 \log (12.6/0.001) \text{ dB} \\
&= 41 \text{ dB}
\end{aligned}
$$

The magnitude of L_P may be positive, negative or zero, depending upon whether P is greater than, less than, or equal to P_o.

In some countries, the unit of power level difference is the neper (symbol Np). The magnitude of the power level difference is then expressed by

$$L_P = \{0.5 \ln (P/P_o)\}\,[\text{neper}] \tag{2}$$

Note that the numerical value is expressed in terms of the natural logarithm of the power ratio.

From equations (1) and (2) we can readily deduce that the [neper] is larger than the [decibel], the relationship being

$$1 \text{ [Np]} = 20/(\ln 10) \text{ [dB]}$$
$$1 \text{ [Np]} = 8.685\,890 \text{ [dB]}$$

Thus, a power level difference of 41 dB is the same as a power level difference of 4.72 Np.

Amplitude level difference

Amplitude level difference (symbol L_F) is a quantity whose magnitude is related logarithmically to the amplitude ratio. The unit of amplitude level difference is also the decibel. However, the magnitude of L_F is given by

$$L_F = \{20 \log (F/F_o)\} \text{ [decibels]} \tag{3}$$

Note that the numerical value in equation (3), is different from that in equation (1) because the factor 10 is replaced by the factor 20.

For example, if the reference voltage across a load is 3 V while the actual voltage is 0.7 V, the amplitude level difference L_E is

$$
\begin{aligned}
L_E &= 20 \log (E/E_o) \text{ [dB]} \\
&= 20 \log (0.7/3) \text{ [dB]} \\
&= -12.64 \text{ dB}
\end{aligned}
$$

Whenever the nature of the amplitude is known (voltage, current, pressure, etc.) the amplitude level difference L_F is expressed in terms of voltage, current, pressure, etc. Thus, in the above example, L_E is called the *voltage* level difference. Similarly, in acoustics, we speak of a *pressure* level difference.

The neper is also used as a unit of amplitude level difference. In this case, the magnitude of the amplitude level difference L_F is given by

$$L_F = \ln (F/F_o) \text{ [neper]} \tag{4}$$

The amplitude relationship between the neper and the decibel remains the same as the power relationship: $1 \text{ [Np]} = 8.685\,890 \text{ [dB]}$

Relationship between L_P and L_F

In comparing equations (1) and (3) one may ask why different factors (10 and 20) were chosen for L_P and L_F. The reason is that in many practical cases, the square of the amplitude of a voltage, current or pressure can be used as a proxy for the power at the point of measurement.

For example, in electrotechnology, for a fixed impedance, the power is proportional to the square of the voltage E. This enables us to express the power level difference L_P in terms of the voltage level difference L_E as follows:

$$
\begin{aligned}
L_P &= 10 \log (P/P_o) \\
&= 10 \log (E/E_o)^2 \\
&= 20 \log (E/E_o) \\
&= L_E
\end{aligned}
$$

Thus, when the impedance is fixed $L_P = L_E$. In other words, the amplitude level difference has the same numerical value as the power level difference.

It is usually much easier to measure a voltage, current or pressure than the actual power, and so the concept of amplitude level difference L_F is useful in the determination of L_P.

Standardized amplitude levels and power levels

In science and engineering, it has been found convenient to establish a few standardized amplitudes and powers and use them as reference levels. For example, the international standard for the sound power radiated by a source has been standardized at 1 picowatt. Thus, $P_o = P$ (ref 1 picowatt) $= 10^{-12}$ W.

Similarly, the reference sound pressure has been standardized at 20 micropascals. These standards enable us to express a given acoustic power or a given sound pressure as an absolute sound power level or sound pressure level.

Example:

Calculate the absolute sound pressure level for a sound having an acoustic pressure of 1.2 Pa.

Solution:

$$
\begin{aligned}
L_F &= 20 \log F/F \text{ (ref 20 } \mu\text{Pa)} \\
&= 20 \log (1.2/20 \times 10^{-6}) \\
&= 95.56 \text{ dB (ref 20 } \mu\text{Pa)}
\end{aligned}
$$

In electrical work, the standardized reference power is 1 mW. Consequently, a power of 10 kW corresponds to an absolute power level of

$$
\begin{aligned}
L_P &= 10 \log P/P \text{ (ref 1mW)} \\
&= 10 \log (10^4/10^{-3}) \\
&= 70 \text{ dB (ref 1 mW)}
\end{aligned}
$$

When giving power or pressure levels it is important to state the reference power or pressure that was used in making the calculation.

APPENDIX IV

QUANTITIES AND UNITS

Quantities

A quantity is any physical property that can be measured. We use about 400 quantities to describe and measure the physical world around us. A few of these quantities are listed below.

length	work	area	electromotive force
time	energy	density	entropy
mass	speed	angle	pressure
force	acceleration	volume	momentum
torque	power	luminance	viscosity

A quantity may also be a physical constant, such as the gas constant, the Planck constant and the rest mass of an electron.

Relationship between Quantities

The study of physics and mathematics has shown that all quantities are interdependent and so it is always possible to relate them in some way. The relationship can be established by definition, by geometry, by physical law, or by a combination of these three.

Pressure, for example, is a quantity which is related, by definition, to a quantity *force* divided by a quantity *area*. *Area*, in turn, is a quantity related, by geometry, to the product of two quantities of length. *Force*, on the other hand, is a quantity related, by Newton's second law, to the quantity *mass* times the quantity *acceleration*.

Even a seemingly isolated quantity such as *temperature* is related to the quantities *pressure*, *volume* and *mass* by virtue of the behaviour of gases. We can even relate the quantities *length* and *time* by using that universal constant, the speed of light. Thus, if we define our concepts correctly we can relate any quantity to any other by one or more of the three ways mentioned above.

The relationships between quantities are expressed in the form of *quantity equations*. Thus, the equation *area = length × width* is a quantity equation which states that the quantity (area of a rectangle) is equal to the quantity (length) times the quantity (width).

Base Quantities

To deduce a set of quantity equations we must first establish a number of so-called base quantities. Base quantities are the building blocks upon which we erect the entire structure and relationships of the physical world. The number of base quantities, as well as their choice, is quite arbitrary but, in general, we try to select quantities that are easy to understand, that are frequently used and for which accurate, measurable standards can be set.

In this context, the International System of Units, or SI, makes use of seven base quantities: *mass, length, time, temperature, electric current, luminous intensity* and *amount of substance*.

Derived Quantities

Derived quantities are those which can be deduced by definition, by geometry or by physical law, using the selected base quantities as building blocks. Examples of derived quantities are *velocity* (length/time), *area* (product of two lengths) and *force* (mass × acceleration).

Units

A unit is a quantity that has a definite, defined magnitude. For example, the unit of time we call second is a quantity whose magnitude is defined as follows: "One second is the duration of 9 192 631 770 periods of radiation corresponding to the transition between the two hyperfine levels of the ground state of the cesium - 100 atom".

A unit, therefore, is not just a convenient name, but represents a quantity (time, force, viscosity, etc.) having a precise, stated magnitude. Because units have a definite size, they can be used as measuring sticks to specify the magnitude of a quantity. Clearly, the unit must be of the same nature as the quantity being measured.

As measurement systems evolved, different units were devised, — often to satisfy the needs of a particular specialty in science or engineering. For example, units of pressure such as the bar, inch of mercury, atmosphere and pound per square inch were introduced. The International System of Units was especially created to eliminate the need for such a broad variety of units. In effect, the SI establishes a unique set of units to measure the approximately 400 quantities of practical interest.

Owing to the tremendous range of magnitudes that a quantity may have, it is useful to establish multiples and submultiples of the unit used to measure it. Thus, the kilometer is a unit whose length is precisely 1000 times the defined length of the meter. Similarly, the micrometer is a unit whose length is precisely one millionth of the defined length of the meter.

Magnitude of a quantity

One of the important characteristics of a quantity is its magnitude. Thus, a quantity such as area can extend from the cross section of an atomic nucleus to the size of a football field. Similarly, a quantity like energy can range from that of an orbiting electron to the energy contained in a ton of coal.

As stated previously, the magnitude of any quantity is measured by comparing it with the size of some convenient *unit*. The length of a ship, for example, can be stated by comparing it with the length of the [foot]. The energy in a pile of coal can likewise be stated by comparing it with a unit of energy, such as the [Btu].

The brackets [ ] around "foot" draw our attention to the fact that we are referring to the size of a unit of length, and not simply to some convenient name.

However, the length of the ship could be measured equally well by comparing it with the [yard], and the energy in coal could be specified by comparison with the [joule]. Obviously, the unit we use does not affect the magnitude of the quantity we are measuring, but does affect the numerical value $\hat{N}$ defined below.

The magnitude of a quantity is defined by the equation

$$\text{magnitude of a quantity} = \hat{N} \times [U]$$

in which [U] is the unit and $\hat{N}$ is the number of times the unit must be taken to make up the quantity. $\hat{N}$ is therefore a pure number.

Thus, when we state that a ship is 180 feet long, we are saying, in effect, that

$$\text{magnitude of the length} = 180 \times [\text{foot}]$$

Similarly, when an electric motor develops 400 horsepower we mean

$$\text{magnitude of the power} = 400 \times [\text{horsepower}]$$

which is to say the motor produces 400 times as much power as that of a [horsepower] unit.

These simple statements concerning units and quantities will lead us to some useful conclusions when dealing with equations.

Base Units

For every base quantity in a measurement system, there exists a corresponding base unit. Thus, in the SI, the *kilogram* (kg), *meter* (m), *second* (s), *kelvin* (K), *ampere* (A), *mole* (mol) and *candela* (cd) are the base units (and their symbols) which correspond respectively to the base quantities mass, length, time, temperature, electric current, amount of substance and luminous intensity.

Because base units are the fundamental measuring sticks, it is essential that their magnitude be precisely defined. These definitions are given in Appendix II.

Derived Units

Every derived quantity has a corresponding derived unit. Some of these units are expressed simply as products or as ratios of the base units of which they are composed. This is the case for the derived unit of area (m^2) and for the derived unit of velocity (m/s).

However, it is sometimes useful to give a special name to a derived unit, particularly when its expression in terms of base units is cumbersome. For example, the derived unit of energy in the SI is the kilogram meter squared per second squared ($kg \cdot m^2/s^2$); but it has been given the much shorter name, the joule, (J).

Dimension of Derived Units

Derived units are said to possess a *dimension*. The dimension of a derived unit is its value in terms of base units. For example, the base dimension of the [newton] (symbol N) is [$kg \cdot m/s^2$]. Thus, 1 N = 1 $kg \cdot m/s^2$.

The SI dimensions of the units of force, power, voltage, magnetic flux, etc., are displayed between brackets on the left-hand side of each conversion chart. Some have surprising dimensions, leading us to think we may have discovered the very essence of truth. The volt, for example, has the dimension [$kg \cdot m^2/A \cdot s^3$] which means exactly what it says – one volt is equal to one kilogram meter squared per ampere second cubed. But lest we attach some deep meaning to this dimension, we should remember that it all depends upon the quantity equations, and how the base units were defined.

Standards

All base units are essentially *definitions* of physical quantities, and so they are always exact. Base units are measured in the real world by *standards* which are the physical embodiment of the base units. These standards can approach the defined values of the base units very closely, but because the standards are affected by temperature, pressure and other extraneous variables, they measure the base unit with high accuracy only under prescribed conditions. Typical accuracies that are attainable in standards laboratories for SI base units are listed below.

kilogram:	1 part in 10^8	ampere:	1 part in 10^6
meter:	1 part in 10^{10}	kelvin:	1 part in 10^6
second:	1 part in 10^{13}	candela:	5 parts in 10^3
		mole:	1 part in 10^{23}

APPENDIX V

QUANTITY EQUATIONS AND
NUMERICAL EQUATIONS

In dealing with scientific relationships it is important to distinguish between two types of equations: quantity equations and numerical equations. Both types are encountered in texts and reference books and the concept of units and quantities is useful in understanding their respective attributes.

Quantity Equations

The energy in a hurricane, the pressure at the bottom of the sea, the weight of a stone and the viscosity of oil are quantities of nature which existed long before man was on earth to measure them. And, whether man measures them or not, these quantities are there, interacting with each other according to fundamental laws. Because quantities — and not units — conform to the laws of nature, physicists often express these laws in terms of *quantity equations.**

Quantity equations possess two important attributes:

1. They show the relationship between quantities.

2. They can be used with any system of units.

There are three basic types of quantity equations:

1. Quantity equations established from the laws of nature; example:

    Newton's second law of motion

    $$F = ma$$

    in which

    F = magnitude of the force
    m = magnitude of the mass
    a = magnitude of the acceleration

2. Quantity equations established from a definition; example:

    definition of pressure

    $$p = F/A$$

* quantity equations are also called physical equations, or equations between quantities.

in which

p = magnitude of the pressure
F = magnitude of the force
A = magnitude of the area

3. Quantity equations arising from geometry; example:

Area of a circle

$$A = \pi r^2$$

in which

A = magnitude of the area
r = magnitude of the radius
π = coefficient based upon the geometry of a circle.

None of these equations imposes a particular set of units. Consequently, we are free to use any convenient units to describe the magnitudes of the quantities F, m, a, p, A and r.

Many quantity equations are a combination of the three basic types mentioned above, but in all cases we can use any units we please to describe the magnitudes of the quantities that are involved.

Some quantity equations contain a physical constant, such as the constant R_a in the equation giving the properties of dry air:

$$pV = R_a mT$$

in which

p = magnitude of the pressure
V = magnitude of the volume
m = magnitude of the mass (of dry air)
T = magnitude of the absolute temperature
R_a = magnitude of the constant for dry air = 287 J/(kg·K)

The magnitude of R_a is expressed in SI units, but this places no restriction whatever on the units that may be used for the other quantities p, V, m and T.

The freedom to use any system or *combination* of units is the reason why quantity equations are so useful. Their elegant simplicity also reflects the basically simple laws of nature and the orderly structure of geometry. The following example illustrates the versatility of these equations in accommodating any set of units.

Example:

Calculate the value of the earth's centrifugal force which acts on a 150-pound man located at the equator. The radius of the earth is 4000 miles and its angular velocity is 1 revolution/day.

Solution:

The quantity equation for centrifugal force is

$$F = m\omega^2 r$$

in which F is the force, m the mass, ω the angular velocity and r the radius (expressed in any units at all).

$$
\begin{aligned}
F &= m\omega^2 r \\
&= 150 \text{ lb} \times (1 \text{ r/d})^2 \times 4000 \text{ mi} \\
&= 600\,000 \; \frac{\text{lb} \cdot \text{r}^2 \cdot \text{mi}}{\text{d}^2}
\end{aligned}
$$

$$(r = \text{revolution} ; d = \text{day})$$

Referring to the appropriate charts (page 30, 9, 25, 42, 24) in order to convert all units to SI base units we find:

$$F = 600\,000 \; \frac{\text{kg}}{2.204\,62} \times \frac{(4 \times 1.570\,80)^2}{(24 \times 3600 \text{ s})^2} \times (1.609\,344 \times 1000) \text{ m}$$

$$
\begin{aligned}
&= 2.315 \text{ kg} \cdot \text{m/s}^2 \\
&= 2.315 \text{ N}
\end{aligned}
$$

The centrifugal force is 2.315 N, or 0.520 lbf (pound-force).

Numerical Equations

We have seen that quantity equations can be used with any system of units. However, it is sometimes convenient to set up an equation to accommodate a *particular* set of units. Such an equation is called a *numerical equation** because all its algebraic terms stand for pure numbers.

We can readily transform a quantity equation into a numerical equation (1) by specifying the units we wish to employ, and (2) by using the following procedure.

1. The quantities Q_1, Q_2, ... Q_n in the quantity equation are respectively replaced by the terms $\hat{N}_1[U_1]$, $\hat{N}_2[U_2]$, ... $\hat{N}_n[U_n]$, where $[U_1]$, $[U_2]$ ... $[U_n]$ are the respective units we wish to employ, and the symbols $\hat{N}_1, \hat{N}_2 \ldots \hat{N}_n$ represent pure numbers.

---

* numerical equations are also called equations between numerical values.

2.   The resulting equation is simplified by grouping the units $[U_1, [U_2] \ldots [U_n]$ together. Each unit is then expressed in terms of SI base units. It will be found that the base units all cancel, leaving us with a pure number $\hat{k}$ whose value depends upon the units $[U_1], [U_2], \ldots [U_n]$ that were originally selected.

3.   The final numerical equation is therefore composed of the symbols $\hat{N}_1$, $\hat{N}_2$, $\ldots \hat{N}_n$ (each of which represents a pure number) and the pure number $\hat{k}$.

In referring to technical literature, how can we tell the difference between a quantity equation and a numerical equation? In a numerical equation, each algebraic term represents the *numerical* value of the quantity and specifies the unit that applies thereto. On the other hand, in a quantity equation, each algebraic term represents only the magnitude of each quantity and no units are specified. The following example illustrates the difference between the two, using the equation for the centrifugal force that acts on a revolving body.

Quantity Equation for centrifugal force

$$F = m\omega^2 r$$

in which

$F =$ centrifugal force

$m =$ mass

$\omega =$ angular velocity

$r =$ length of radius

Numerical Equation for centrifugal force

$$\hat{F} = 0.003342 \; \hat{m}\hat{\omega}^2\hat{r}$$

in which

$\hat{F} =$ centrifugal force, in newtons

$\hat{m} =$ mass, in kilograms

$\hat{\omega} =$ angular velocity, in revolutions per minute

$\hat{r} =$ length of radius, in feet

$0.003342 =$ numerical coefficient

In the above quantity equation any units may be used, but in the numerical equation the mass must be in kilograms, the angular velocity must be in revolutions per minute and the radius must be in feet. Furthermore, the resulting centrifugal force is expressed in newtons.

How is the numerical equation arrived at? It is derived from the quantity equation using the procedure described above. The detailed methodology will now be given, knowing that we want to express F in newtons, m, in kilograms, ω in revolutions per minute and r in feet. (In order to clearly distinguish between algebraic symbols that represent quantities from those that represent numerical values, we have put a "hat" on the latter).

First, we have the quantity equation

$$F = m\omega^2 r$$

but

$$F = \text{(number of force units)} \times \text{[desired force unit]}$$
$$= \hat{N}_1[U_1]$$
$$= \hat{F}\ [N]$$

and

$$m = \text{(number of mass units)} \times \text{[desired mass unit]}$$
$$= \hat{N}_2[U_2]$$
$$= \hat{m}\ [kg]$$
$$\omega = \text{(number of angular velocity units)} \times \text{[desired angular velocity unit]}$$
$$= \hat{N}_3[U_3]$$
$$= \hat{\omega}\ \text{[revolution per minute]}$$
$$= \hat{\omega}\ \text{[r/min]}$$
$$r = \text{(number of length units)} \times \text{[desired length unit]}$$
$$= \hat{N}_4[U_4]$$
$$= \hat{r}\ [ft]$$

Substituting the magnitude of these quantities in the quantity equation, we obtain

$$F = m\omega^2 r$$
$$\hat{F}\ [N] = \hat{m}\ [kg]\ (\hat{\omega}\ [r/min])^2\ \hat{r}\ [ft]$$

grouping the units together, we obtain

$$\hat{F} = \hat{m}\hat{\omega}^2\hat{r}\ [kg]\ [r/min]^2\ [ft]\ [N^{-1}]$$

then, expressing the units in terms of SI base units

$$\hat{F} = \hat{m}\hat{\omega}^2\hat{r}\ [kg]\ [2\pi/60s]^2\ [m/(1.093\ 61 \times 3)]\ [N^{-1}]$$

$$= \hat{m}\hat{\omega}^2\hat{r}\ \frac{(2\pi/60)^2}{3.281}[kg\cdot m/s^2]\ [N^{-1}]$$

Finally, recalling that $1\ N = kg\cdot m/s^2$, we see that all the base units cancel and so we obtain the resulting numerical equation:

$$\hat{F} = 0.003342\ \hat{m}\hat{w}^2\hat{r}$$

The above derivation shows clearly that the algebraic symbols $\hat{F}, \hat{m}, \hat{\omega}, \hat{r}$ represent numerical values and the coefficient $\hat{k} = 0.003342$ is also a pure number.

It is obvious that we can generate as many numerical equations as we please from a given quantity equation. The number is limited only by the variety of units we wish to employ. Furthermore, whenever the units are changed, the value of the coefficient k changes.

Meaning of Coherence

A system of units is said to be coherent when the numerical equations, expressed in base units, have the same numerical factors as the corresponding quantity equations. Coherence is one of the distinguishing features of the SI. The following list of equations illustrates the meaning of coherence.

Quantity Equation	**Numerical Equation**
$W = Fd$	$\hat{W} = \hat{F}\hat{d}$
$F = ma$	$\hat{F} = \hat{m}\hat{a}$
$E = 1/2\ mv^2$	$\hat{E} = 1/2\ \hat{m}\hat{v}^2$
$F = G\ m_1m_2/d^2$	$\hat{F} = 6.673 \times 10^{-11}\ \hat{m}_1\hat{m}_2/\hat{d}^2$

in which

in which

$W =$ work	$\hat{W} =$ work in joules
$F =$ force	$\hat{F} =$ force in newtons
$d =$ distance	$\hat{d} =$ distance in meters
$E =$ energy	$\hat{E} =$ energy in joules
$m =$ mass	$\hat{m} =$ mass in kilograms
$v =$ velocity	$\hat{v} =$ velocity in meters per second
$a =$ acceleration	$\hat{a} =$ acceleration in meters per second squared
$G = 6.673 \times 10^{-11}\ m^3{\cdot}kg^{-1}{\cdot}s^{-2}$.	

Note that the coefficient G is composed of a numerical factor and base units.

Note that the numerical factor 6.673×10^{-11} is the same as in the quantity equation.

Converting a numerical equation into a quantity equation

A numerical equation can readily be converted into a quantity equation, thus rendering it universal as far as units are concerned. The procedure, described below, is the reverse of that used to convert a quantity equation into a numerical equation:

1. The terms $\hat{N}_1, \hat{N}_2 \ldots \hat{N}_n$ in the numerical equation are replaced by the terms $Q_1/[U_1], Q_2/[U_2], \ldots Q_n/[U_n]$, where $Q_1, Q_2, \ldots Q_n$ represent the respec-

tive quantities, and $[U_1]$, $[U_2]$ . . . $[U_n]$ are the corresponding units specified in the numerical equation.

2. The resulting equation is then simplified by grouping the units $[U_1]$, $[U_2]$ . . . $[U_n]$ together, and expressing each in terms of SI base units. The units are manipulated as if they were algebraic symbols. Consequently, they can be reduced to their most simple algebraic expression whenever this is convenient.

3. The resulting quantity equation is composed of the symbols Q_1, Q_2, . . . Q_n and a coefficient K consisting of a pure number and one or more base units. In many important cases, the coefficient has the dimensionless value of 1.

As an example of such a conversion, consider the following numerical equation giving the fusing current of a copper conductor in relation to its diameter.

$$\hat{I} = 70 \; \hat{d}^{1.5}$$

in which

$\hat{I} =$ current, in amperes (in this case, $\hat{I}$ corresponds to $\hat{N}_1$)
$\hat{d} =$ diameter, in millimeters (in this case, $\hat{d}$ corresponds to $\hat{N}_2$)

This equation is typical of many empirical equations in which the relationships are determined experimentally.

First, we know that,

magnitude of current $= I =$ (number of current units) $\times$ [given current unit]

$$= \hat{N}_1[U_1]$$
$$= \hat{I} \, [A]$$

therefore

$$\hat{I} = I/[A]$$

Furthermore,

magnitude of diameter $= d =$ (number of length units) $\times$ [given length unit]

$$= \hat{d} \, [mm]$$

therefore

$$\hat{d} = d/[mm]$$

Referring to the numerical equation, if we now substitute the numerical values $\hat{I}$ and $\hat{d}$ by the respective magnitudes divided by the corresponding unit, we obtain

$$\hat{I} = 70\ \hat{d}^{1.5}$$

$$
\begin{aligned}
I/[A] &= 70\ (d/[\text{mm}])^{1.5} \\
I &= d^{1.5}\ 70\ [A]\ [\text{mm}]^{-1.5} \\
&= d^{1.5}\ 70\ [A]\ [\text{m}/1000]^{-1.5} \\
&= d^{1.5}\ 70\ [A]\ [\text{m}]^{-1.5}\ [1000]^{1.5} \\
&= 2.21 \times 10^{6}\ [A]\ [\text{m}]^{-1.5}\ d^{1.5}
\end{aligned}
$$

Thus, we obtain the quantity equation

$$I = k\ d^{1.5}$$

in which

$I =$ magnitude of the current

$d =$ magnitude of the diameter

$k =$ constant whose magnitude is $2.21 \times 10^{6}\ \text{A·m}^{-1.5}$

This example also shows that units can be treated like algebraic symbols, as far as exponents are concerned.

Discovering Dimensions

To determine the dimension of a derived unit we must know two things:

1. The definition of the derived unit,

2. The quantity equation which connects the derived unit with units whose dimensions are already known.

Let us determine, for example, the dimension of the newton. The SI definition states that "the newton is that force which gives to a mass of one kilogram an acceleration of one meter per second per second".

The quantity equation is

$$F = ma$$

Based upon the definition, we can write:

$$
\begin{aligned}
\text{one [newton]} &= \text{one[kilogram]} \times \text{one [meter per second per second]} \\
1\ [N] &= 1\ [\text{kg}] \times 1\ [\text{m/s}^2] \\
\text{whence } N &= \text{kg·m·s}^{-2}
\end{aligned}
$$

Knowing the dimension of the newton, we can now determine the dimension of the joule. By definition, "the joule is the work done when the point of application of one newton is displaced a distance of one meter in the direction of the force".

The quantity equation is

$$E = Fd$$

that is,   work = force × distance

Using the definition and the quantity equation, we can therefore write:

one [joule] = one [newton] × one [meter]
1 [J] = 1 [N] × 1 [m]

thus

$$J = N \cdot m$$

but since

$$N = kg \cdot m \cdot s^{-2}$$

we obtain

$$J = kg \cdot m^2 \cdot s^{-2}$$

and so the dimension of the joule is $kg \cdot m^2 \cdot s^{-2}$

We are now in position to determine the dimension of the watt. By definition, "the watt is the power which gives rise to the production of energy at the rate of one joule per second. The quantity equation is

$$P = E/t$$

that is, power = $\dfrac{\text{energy}}{\text{time}}$

Using the definition, we can therefore write

one [watt] = $\dfrac{\text{one [joule]}}{\text{one [second]}}$

1 [W] = $\dfrac{1 \text{ [J]}}{1 \text{ [s]}}$

therefore $W = J \cdot s^{-1}$

but since $J = kg \cdot m^2 \cdot s^{-2}$

we obtain $W = kg \cdot m^2 \cdot s^{-3}$

and so the dimension of the watt is $kg \cdot m^2 \cdot s^{-3}$.

Let us now discover the dimension of the volt. By definition "one volt is the difference of potential between two points of a conducting wire carrying a constant current of one ampere, when the power dissipated between these points is equal to one watt".

The quantity equation is

$$U = P/I$$

that is,

$$\text{electromotive force} = \frac{\text{power}}{\text{current}}$$

Using the definition in the quantity equation, we can therefore write:

$$\text{one [volt]} = \frac{\text{one [watt]}}{\text{one [ampere]}}$$

$$1 \text{ [V]} = \frac{1 \text{ [W]}}{1 \text{ [A]}}$$

therefore $V = W \cdot A^{-1}$

but since $W = kg \cdot m^2 \cdot s^{-3}$

we obtain $V = kg \cdot m^2 \cdot A^{-1} \cdot s^{-3}$

and so the dimension of the volt is $kg \cdot m^2 \cdot A^{-1} \cdot s^{-3}$

The dimensions of derived units are found, therefore, by a pyramiding process in which each successive derived unit becomes the stepping-stone to establish the next. That is how the dimension of each quantity listed in the charts was established. It is, however, important to remember that these dimensions are all derived from the base units of the SI system of measurement.

BIBLIOGRAPHY

Ambler, E. 1971. *SI Units, Philosophical Basis for the Base Units.* Technical News Bulletin, March 1971, National Bureau of Standards, Washington, DC.

ANSI/IEEE, 1982. *American National Standard Metric Practice.* (ANSI/IEEE Std 268-1982) New York : Institute of Electrical and Electronics Engineers.

BIPM, 1985. *The International System of Units (SI).* Sèvres, France. Bureau International des Poids et Mesures.

Cohen, E.R. and Taylor B.N. 1986. *The 1986 Adjustment of the Fundamental Physical Constants.* Codata Bulletin. New York. Pergamon.

CSA, 1979. *Canadian Metric Practice Guide. CAN3-Z234.1-79.* Toronto. Canadian Standards Association.

Drazil, J.V. 1983. *Quantities and Units of Measurement.* London. Mansell Publishing.

Engrand, J.C. 1981. *Units and Their Equivalences.* Paris. Vuibert.

Horvath, A.L. 1987. *Conversion Tables of Units for Science and Engineering.* New York. Elsevier.

Huntley, H.E. 1967. *Dimensional Analysis.* New York. Dover Publications.

International Electrotechnical Commission, 1981. *IEC - Publication 27.* Geneva. IEC.

International Commission of Radiation Quantities and Units, 1980. *Radiation Quantities and Units, ICRU Report 33.* Washington. ICRU.

International Organization for Standardization, 1981. *ISO - 31 Series.* Geneva. ISO.

Ipsen, D.C. 1960. *Units, Dimensions and Dimensionless Numbers.* New York, McGraw-Hill.

Judson, L.V. 1976. *Weights and Measures Standards of the United States: A Brief History.* National Bureau of Standards publication 447.

Kerwin, L. 1978. International Union of Pure and Applied Physics. *Symbols, Units and Nomenclature in Physics.* Québec. Université Laval.

McGlashan, M.L. 1973. *Manual of Symbols and Terminology for Physico-Chemical Quantities and Units.* International Union of Pure and Applied Chemistry (IUPAC). New York, Pergamon.

Martinek, A. *Metric System (SI) in Engineering Technology.* Waterloo. Reeve Bean.

Maxwell, J.C. 1954. *A Treatise on Electricity and Magnetism,* Vol. 2 Chap X. New York. Dover Publications.

Preston-Thomas, H. et al. 1968. *An Absolute Measurement of the Acceleration Due to Gravity*. NRC Publication 5693. Ottawa. National Research Council of Canada.

Wildi, T. 1970. *Units*. Québec. Volta.
———— 1972. *Understanding Units*. Québec. Volta.

Young, L. 1969. *Systems of Units in Electricity and Magnetism*. Edinburgh. Oliver and Boyd.

INDEX

A

abampere, 17
abcoulomb, 17
abfarad, 15
abhenry, 15
abmho, 15
abohm, 15
abvolt, 17
acceleration, 9
accuracy, 4
acre, 12
acre-foot, 46
amount of substance, 10
ampere, 17, 51
    per meter, 29
    per square meter, 16
    per square centimeter, 16
    per square inch, 16
ampere-hour, 17
ampere-turn, 29
    per inch, 29
    per meter, 29
amplitude level difference, 55, 56
angle, 9
    radian, 9, 52
    right, 9
    steradian, 52
Angstrom, 25
angular velocity, 11
apostilb, 27
are, 12
area, 12
arpent (French measure of area), 12
arpent (French measure of length), 25
assay ton (long), 30
assay ton (short), 30
atomic mass unit, 31, 49
astronomical unit, 25
atmosphere, 37
atmosphere, technical, 37
atto, 8

B

bar, 37
barn, 12
barrel (UK), 46
    oil (US), 46
barye, 37
base unit (SI), 50
bel, 36
becquerel, 38, 52
biot, 17
board foot, 47
boiler horsepower, 34
Bohr magneton, 28
Bohr radius, 26
Boltzmann constant, 49
British thermal unit, 19
    (International Steam Table), 21
    (ISO/TC 12), 21
    (mean), 21
    (thermochemical), 21
    (39°F), 21
    (60°F), 21
    per hour, 34
    per minute, 34
    per second, 34
    per (hour) ($foot^2$) (°F per foot of thickness), 41
    per square foot hour, 41
Btu (British thermal unit), 19
bushel (CAN, UK, US), 47

C

caliber, 25
calorie, 19
    dietetic, 20
    (International Steam Table), 20
    (mean), 20
    (thermochemical), 20
    (15°C), 20
    (20°C), 20
    per second, 34
    per second centimeter degree Celsius, 41
    per square centimeter second, 35

candela, 27, 52
    per square centimeter, 27
    per square foot, 27
    per square inch, 27
    per square meter, 27
candela steradian, 27
    per square meter, 27
capacitance, 15
capacity, (liquid), 46
    (dry), 47
carat, 30
carbon 12 (atom of), 31
Celsius, 40, 52
cental, 30
centiare, 12
centi, 8
centigrade, 40
centigrade heat unit, 19
centimeter, 25
    (cubic), 46
    (square), 12
    of mercury 60°F, 37
    of water 60°F, 37
    per second, 44
    per second squared, 9
centipoise, 45
centistokes, 45
century, 42
chain, engineer's, surveyor or gunter, 25
    (square), 12
charge, electric, 17
cheval vapeur, 34
CHU, 34
circular frequency, 11
circular mil, 12
coherence, 66
Compton wavelength of the electron, 26
Compton wavelength of the neutron, 26
Compton wavelength of the proton, 26
concentration, 13
conductance, 15
cord, 47
coulomb, 17, 52
    per second, 17

cubic,
    centimeter, 46
    decimeter, 46
    decimeter per second, 23
    foot, 46
    foot per minute, 23
    foot per second, 23
    inch, 46
    meter, 46
    meter per hour 23
    meter per second 23
    mile per year, 23
    yard, 46
    yard per second, 23
cup, (CAN, UK, US), 46
curie, 38
current, electric, 17
    density, 16
cycle per second, 24

D

dalton, 31
day, 42
deca, 8
decade, 42
deci, 8
decibel, 36, 54
decineper, 36, 54
definitions of units, 50-53
degree (angle), 9
    per second, 11
degree (temperature), 40
    Celsius, 40
    Centigrade, 40
    Fahrenheit, 40
    Rankine, 40
    Reaumur, 40
density, 14
derived units (SI), 51, 59
deuteron rest mass, 31
dimension of a unit, 4, 60, 68
dram (mass), 30
    troy, 30
    avoirdupoids, 30

dram (volume), 46
    (CAN,US), 46
dynamic viscosity, 45
dyne, 24
    centimeter (energy), 19
    centimeter (torque), 43
    centimeter per second, 34
    per square centimeter, 37

E

electric field strength, 16
electricity, 15-18
electromotive force, 17
electron charge, 17
electron radius, 26
electron rest mass, 31
electronvolt, 19
elementary entity, 10
energy, 19
erg, 19
    per second, 34

F

Fahrenheit, 40
farad, 15, 52
faraday, 17, 49
fathom, 25
femto, 8
fermi, 25, 26
fifth, (US), 46
flow, 23
fluid ounce, (CAN, UK, US), 46
foot,
    (cubic), 46
    (square), 12
    per minute, 44
    per second, 44
    per second squared, 9
    of water 60°F, 37
french foot, 25
foot pound-force, (pound-force foot),
    (energy), 19
    (torque), 43

per hour, 34
per minute, 34
per second, 34
foot poundal, 19
footcandle 27
footlambert, 27
force, 24
    moment of, (see torque),
franklin, 17
free fall, 9
french foot (square), 12
frequency, 24
furlong, 25

G

Gal (galileo), 9
gallon,
    (CAN, UK, US, dry), 47
    (CAN, UK, US, liquid), 46
    per day (CAN, UK, US), 23
    per minute (CAN, UK, US), 23
    per second (CAN, UK, US), 23
gamma, 29
gauss, 29
GeV, 19
giga, 8
gigahertz, 24
gilbert, 29
gill (CAN, UK, US), 46
grad, 9
grain, 30
gram, 30
    calorie (see calorie), 19
    per cubic centimeter, 14
    per liter, 14
    per milliliter, 14
gram-force, 24
    centimeter, 43
    per square centimeter, 37
gray, 39, 52

H

hectare, 12
hecto, 8
hectobar, 37
henry, 15, 52
hertz, 24, 53
hogshead (US), 46
horsepower, 34
    (550 foot lbf/s), 36
    (boiler), 34
    (electric), 36
    (metric), 36
    (UK), 36
    (water), 36
    hour, 19
hour, 42
hundredweight, 30
hundredweight (long), 30

I

illuminance, 27
inch, 25
    (cubic), 46
    (square), 12
    micro, 25
    of mercury (60°F), 37
    of water (60°F) 37
    per second, 44
inductance, 15
induction, magnetic (see magnetic
                flux density)
International System of Units (SI), 50

J

joule, 19, 53
    per second, 34

K

kelvin, 40, 51
kilo, 8
kilocalorie, 19

kilogram, 30, 50
    per cubic decimeter, 14
    per cubic meter, 14
    per liter 14
kilogram-force, 24
    meter (energy), 19
    meter (torque), 43
    per square centimeter, 37
    per square meter, 37
    per square millimeter, 37
kilometer, 25
    (square), 12
kilopond, 24
kilovolt, 17
    per inch, 16
kilowatt, 34
kilowatthour, 19
kinematic viscosity, 45
kip, 24
    per square inch, 37
knot, 44

L

lambert, 27
langley per second, 35
league, nautical, 25
    statute, 25
legal subdivision, 12
length, 25
light, 27
    (velocity of), 49
    year, 25
line of force, 29
    per square inch, 29
link, engineer's, 25
    surveyor or gunter, 25
liter, 46
    per minute, 23
    per second, 23
long ton, 30
    per cubic yard, 14

lumen, 27, 53
    per square foot, 27
    per square meter, 27
    per steradian, 27
luminance, 27
luminous flux, 27
luminous intensity, 27
lux, 27, 53

M

magnetic field strength, 29
magnetic flux, 29
magnetic flux density, 29
magnetic moment, 28
    of electron, 28
    of deuteron, 28
    of muon, 28
    of neutron, 28
    of proton, 28
magnetism, 29
magnetomotive force, 29
mass, 30
    resistivity, 18
    (standard of), 60
    per unit volume (see density),
maxwell, 29
mega, 8
megaton TNT, 19
megawatt, 34
megohm, 15
meter, 25, 50
    (cubic), 46
    (square) 12
    per minute, 44
    per second, 44
    per second squared, 9
    squared per second, 45
meter kilogram-force,
    (energy), 19
    (torque), 43
    per second, 34

metric
    horsepower, 36
    tonne, 30
MeV, 19
mho, 15
micro, 8
microinch, 25
microfarad, 15
micron, 25
mil, 25
    (circular), 12
    (square), 12
mile, nautical, 25
    per hour, 44
    per hour second, 9
    per minute, 44
    per second, 44
    (square) 12
    statute, 25
millenium, 42
milli, 8
millimeter, 25
    of mercury (0°C), 37
    (square), 12
minim, (CAN, UK, US), 46
minute, (angle), 9
    (time), 42
mole, 10, 52
mole per cubic meter, 13
moment of force, (see torque),
month, 42
multiples of units, (SI), 4, 5, 8
muon rest mass, 31

N

nano, 8
nautical league (Int), 25
nautical mile, 25
nautical,
neper, 36, 54

neutron rest mass, 31
newton, 24, 53
    per coulomb, 16
    second per square meter, 45
    per square meter, 37
    per weber, 29
newton meter,
    (energy), 19
    (torque), 43
    per second, 34
nit, 27
nuclear magneton, 28
numerical equation, 63, 66

O

oersted, 29
    centimeter, 29
ohm, 15, 53
    centimeter, 18
    circular mil per foot, 18
    gram per square meter, 18
    kilogram per square meter, 18
    meter, 18
    mm^2 per meter, 18
    (micro)-inch, 18
    pound per square mile, 18
ounce, 30
    avoirdupoids, 30
    fluid (CAN, UK), 46
    fluid (US), 46
    troy or apothecary, 30
    - force, 24
    - force foot, 43
    - force inch, 43
    per cubic foot, 14
    per cubic inch, 14

P

pace, 25
parsec, 25
pascal, 37,53
peck (CAN, UK, US), 47
pennyweight, 30

perch, 25
    square, 12
perche, 25
petrograd standard, 47
phot, 27
pica, 25
pico, 8
pieze, 37
pint,
    (CAN, UK, US, liquid), 46
    (CAN, UK, US, dry), 47
plane angle (see radian)
Planck constant, 49
point (length), 25
    (mass), 30
poise, 45
poiseuille, 45
potential, electric (see electromotive
    force),
pound, avoirdupois (mass), 30
    per cubic inch, 14
    per cubic foot, 14
    per cubic yard, 14
    per foot second, 45
pound, troy or apothecary (mass), 30
pound-force (avoirdupois), 24
    foot (energy), 19
    foot (torque), 43
    inch (torque), 43
    second per square foot, 45
    per square foot, 37
    per square inch, 37
poundal, 24
    second per square foot, 45
    per square foot, 37
power, 34
    per unit area, 35
power level difference, 54, 56
prefixes, 5, 8
pressure, 37
proton rest mass, 31
psi, 37
pulsatance, 11

Q

quad, 19
quantity, 57
    base, 57
    derived, 58
    equation, 61, 66
quart (CAN, UK, US, liquid), 46
    (CAN, UK, US, dry), 47

R

rad, 39, 9
radian, 9, 52
    per second, 11
Rankine, 40
Reaumur, 40
register ton, 46
rem, 38
rep, 39
resistivity, 18
    mass, 18
    volume, 18
revolution, 9
    per day, 11
    per hour, 11
    per minute, 11
    per second, 11
right angle, 9
rod, 25
roentgen, 39
rood, 12
Rydberg constant, 49

S

scruple, 30
second (angle), 9
    (time), 42, 50
section, 12
short ton, 30
short ton per cubic yard, 14
SI units,
    definitions of, 50-53
    table of, 50, 51

siemens 15, 53
sievert, 38, 53
slug, 30
    per cubic foot, 14
    per foot second, 45
solid angle (steradian), 52
speed, (see Velocity)
square,
    chain, 12
    mil, 12
    centimeter per second, 45
    foot per second, 45
    meter per second, 45
standards, 60
statampere, 17
statcoulomb, 17
statfarad, 15
stathenry, 15
statmho, 15
statohm, 15
statvolt, 17
statute league, 25
statute mile (or mile), 25
steradian, 52
stere, 46
stilb, 27
stokes, 45
stone, 30
stress, 37
submultiple of a unit (SI), 4
supplementary unit (SI), 50
Système International d'Unités, 50

T

tablespoon (CAN, UK, US), 46
teaspoon (CAN, US), 46
technical atmosphere, 37
temperature, 40
tera, 8
tesla, 29, 53
therm, 19

thermal conductivity, 41
thermie, 19
time, 42
ton, 30
    (long), 30
    (short),, 30
    assay, (long) 30
    assay, (short), 30
    nuclear equivalent of TNT, 19
    of refrigeration, 34
    (long) per cubic yard, 14
    (short) per cubic yard, 14
tonne (metric), 30
torque, 43
torr, 37
township, 12
Troy measure, 30

U

units, 58
    base, 50, 59
    definitions, 50-53
    derived, 51, 59
    dimensions of, 60
    (SI), 50
    supplementary, 50
unit pole, 29

V

velocity, 44
    angular, 11
viscosity, dynamic, 45
    kinematic, 45
volt, 17, 53
    per centimeter, 16
    per meter, 16
    per mil, 16
volume, 46, 47
    rate of flow (see flow),
    resistivity, 18

W

watt, 34, 53
    per centimeter^2, 35
    per foot^2, 35
    per inch^2, 35
    per meter kelvin, 41
    per square meter, 35
watthour, 19
wattsecond, 19
water pressure (see pressure),
weber, 29, 53
    per square meter, 29
week, 42

Y

yard, 25
    (cubic), 46
    (square), 12
year, 42